The Ecosphere

Organisms, Habitats, and Disturbances

C. Leland Rodgers
Professor of Biology
and
Chairman, Department of Biology

Rex E. Kerstetter
Assistant Professor of Biology
Furman University

Harper & Row, Publishers
New York, Evanston, San Francisco, London

Sponsoring Editor: Alvin A. Abbott
Project Editor: Duncan Hazard
Designer: June Negrycz
Production Supervisor: Robert A. Pirrung

The Ecosphere:
Organisms, Habitats, and Disturbances

Library of Congress Cataloging in Publication Data
Rodgers, Charles Leland, 1918–
The ecosphere.

1. Ecology. 2. Man—Influence on nature.
I. Kerstetter, Rex E., joint author. II. Title.
QH541.R63 574.5 73–7479
ISBN 0–06–045552–7 (pbk.)

Contents

Preface

During the decade of the 1960s people began to realize the importance of a good environment and became increasingly determined to improve and preserve it. Until that time ecology had been the domain of professionals and a few amateur zealots, and the word *ecology* was not even included in the vocabulary of most people. Because of the new awareness and its attendant publicity, the term *ecology* has escaped from its narrow academic usage and become a household word. Moreover, large numbers of students who have never before been interested in ecology are seeking some formal training in the subject in order to become better-informed and responsible citizens. Their understanding of ecological relationships and techniques will lay the foundation for rehabilitating and maintaining the ecosphere. They do not want to be overburdened with complex interpretations that require an extensive background of specialized training, but they want enough substantive knowledge of the subject to be able to appraise unemotionally the present conditions and to participate in efforts to improve them. This book provides a suitable foundation for their first formal study of ecology.

The purpose of this book is to assess man's impact and potential in the natural environments. In sequence, it reviews some historical highlights of man's changing role in the environment, describes the major terrestrial and aquatic communities of organisms, and deals with environmental problems and their possible solutions. Students should find the topics of this book timely, since such problems as growth of population, exploitation of resources, and pollution of the environment are treated in considerable detail. They are also provided with a thorough treatment of a variety of natural habitats and their inhabitants, knowledge that is invaluable in broadening their awareness and comprehension of natural surroundings.

Among the useful features of this book are the illustrations, glossary, and appendixes. The illustrations are selected from widely differing geographical areas. The glossary includes most of the terms that are introduced in the book and may be unfamiliar to the beginning student. Lists of names and addresses of a number of private and governmental organizations dealing with environmental problems are included in the appendixes. Enumerated by state in another appendix are many protected locations where distinctive habitats or plants and animals may be found.

It is hoped that students will be guided by this book to a greater appreciation of the environment and will become sensitized to some of its abuses. For basic to the work of preserving the ecosphere is the realization that the lifestyle of the individual needs to adjust to the fact that man is *a part of* and not *apart from* his surroundings.

C. LELAND RODGERS
REX E. KERSTETTER

The Ecosphere

Part 1
Man's Past Concern for the Environment

Chapter 1
The Beginnings of Ecology

MAN IN THE ENVIRONMENT

Although there is evidence that man has long had an intuitive feeling for the relationship between organisms and their environments, it is only recently that he has begun to study these interactions in a systematic way. In his primitive existence as a hunter, early man had to understand how his prey lived in order to increase his chances of making a kill; as a farmer, he had to learn what to do to increase the yield of his crops. Having this knowledge meant survival.

As he struggled for survival, early man cared only for the short-term effects of his activities on his immediate environment. If the long-term results threatened his chances for survival, he could always move on to better hunting grounds or to areas with more fertile soil. In other words, exploitation of the environment was permissible for early man—that is, it had little, if any, negative effect on a population because there remained vast uninhabited areas.

Modern man is still concerned with the age-old problem of how to succeed in his environment. The problem is vastly more complicated, however.

THE UNITY OF LIFE AND ENVIRONMENT

The mere mention of life and the environment as distinct entities detracts from the reality of the unity of their relationship. Life is the product of a complex organization of some of the ingredients of the earth; after death these are freed and made available for reuse by other forms of life. Thus, elements are only temporarily entrapped in a body. The movement within and between the living and nonliving components of the system is endless. The reciprocal exchanges emphasize how closely living organisms and the environment are enmeshed.

Besides being composed of environmental substances and being sustained by the environment, organisms discharge **metabolites** into the environment and, as a result, change it. The environment, for its part, influences the growth and development of plants and animals in numerous ways such as subjecting them to variables like light and temperature. If organisms are to survive, they must be able to tolerate extremes within the composite of these environmental variables. Thus, the abundance of organisms representing any particular species today reflects the success of that species in adapting or evolving to a particular set of environmental conditions in the past. This says little or nothing, however, about its potential for success in the future.

HISTORICAL EVENTS

Despite man's long history of observation of life in natural settings and a general awareness of natural relationships, it was not until 1868 that Reiter coined the name *oekologie* for this important science. The word is derived from two Greek words meaning "house" and "discourse," and essentially means a study of the surroundings of organisms—or, better still, a study of environments. The original word *oekologie* has become *ecology,* meaning the science dealing with relationships between organisms and their environments.

Long before ecology had a name, accounts of natural phenomena, simple and sometimes inaccurate though they were, were left by the Greeks. Their society was undergirded by a system of slavery that freed some people for observation and creative thinking. Hippocrates (460–370 B.C.), known as the father of medicine, used the title "On Airs, Waters and Places" in writing about environmental effects on health. Later, Aristotle (384–322 B.C.) classified animals according to where they lived. His student Theophrastus wrote about communities of plants and color adaptations of animals to their environments. But the study begun by the Greeks stagnated during the Middle Ages, which lasted a thousand years.

With the renaissance of spirit that terminated the Middle Ages, science began to unfold. By the eighteenth century the foundation of "modern" science was beginning to take shape. Advances in physics and chemistry permitted ecologists to increase their understanding of the environment and the requirements of organisms living therein. For example, the discovery of oxygen (1774) by Joseph Priestley and his demonstration that animals remove it from the air and plants put it back again were breakthroughs of fundamental importance to understanding how truly inseparable organisms are from their environments. Shortly thereafter (1798), the Dutch physician Jan Ingenhousz showed that carbon dioxide consumption and oxygen discharge by green plants occur only in the presence of light, another primary environmental factor.

Although the discoveries of ecological importance during and after the eighteenth century are too numerous to mention, we mention some landmark advancements. During the eighteenth and nineteenth centuries, Europeans were actively exploring the earth, a necessary step toward gaining an enlarged perspective of the world and the interrelationships of its biotic and environmental components. Some of the better-known explorers were Joseph Banks (worldwide), Robert Brown (Australia), Alfred R. Wallace (Malayan archipelago), Charles Darwin (worldwide), Joseph D. Hooker (Antarctica, Tasmania, and New Zealand), Alexander von Humboldt (South America), Louis Agassiz (United States and Brazil), Henry W. Bates (South America), and Edward Forbes (Mediterranean region).

Two of the explorers, Wallace and Darwin, synthesized their observations into the now-famous concept of evolution (1858), in which natural selection is a key ingredient. By natural selection they meant that environmental factors, especially a limited food supply, eliminate all but the successful competitors, usually those best adapted to the prevailing conditions. This is clearly an ecological, environment-organism relationship. They were also convinced that organisms inherently overreproduce. Their ideas, at least in part, came from Thomas R. Malthus, who as early as 1798 said that populations tend to increase in geometric proportion until the necessities of life cannot meet the needs of the population, resulting in a struggle for survival. Even before Malthus, some observers had theorized about populations tending to increase in geometric proportion or being limited by disease or want.

P. F. Verhulst (1838) plotted the relationship of population growth to population density and showed it as an *S* curve (then called a logistic curve). It was a clear way to express quantitatively both the potential rate of increase and environmental resistance. Raymond Pearl and L. J. Reed rediscovered Verhulst's *S* curve and its use in 1920. It is discussed thoroughly in Chapter 11 of this book.

The recognition that environmental variables impose limitations on processes or populations has been articulated in various ways by different people. In 1840 Justus von Liebig said that any environmental factor approaching a critical minimum would be a limiting factor. This observation is now referred to as Liebig's **Law of the Minimum**. In 1913 Victor E. Shelford extended Liebig's idea of limitations caused by scarcities to include those caused by excesses such as too much heat or light. Thus, an organism or population can survive between the extremes of scarcity and excess but will die if any environmental factor goes beyond these extremes. This principle is known as Shelford's **Law of Tolerance**.

The study of the composition and interactions of mixed groups of organisms, called community ecology, goes back at least to the Greeks. It received renewed and more comprehensive attention in the nineteenth century—particularly in 1838 with the work of August Grisebach, a German botanist—and is of even greater interest today. Among the noteworthy successors of Grisebach who did community studies were Karl Mobius (oyster beds), S. A. Forbes (lake community), J. E. B. Warming (plant communities), F. E. Clements (plant communities), C. C. Adams (animal communities), V. E. Shelford (animal communities), and Charles Elton (community dynamics).

Around the turn of the century, ecologists began to take an interest in community **succession** (see Chapter 3). Following the lead of J. S. DeLue (1806) and several other Europeans toward the end of the nineteenth century, H. C. Cowles published a classic paper (1899) about succession on a sand dune region bordering Lake Michigan. His work was soon followed by another classic, V. E. Shelford's study on the succession of tiger beetles, published in 1907. At about the same time (1906 and 1909), C. C. Adams completed important successional studies in animal communities of northern Michigan and Isle Royale in Lake Superior. Since then, successional studies have been numerous.

Interestingly, the turn of the century also marks an important milestone in the evolution of ecology as a science, at least for Americans. The new country, which squandered its resources as if they were limitless, bore ugly scars of waste and depletion. The forests were in shambles. Responsible people saw what was happening and initiated a movement for conservation and restoration.

The important step in conservation was made in 1898, when Dr. Carl A. Schenck founded the first American forestry school on the Biltmore Estate (now part of the Pisgah National Forest in North Carolina). He and his predecessor, Gifford Pinchot, saw the devastated woodlands and recognized an opportunity to improve forestry practices. When Pinchot left Biltmore to serve under Theodore Roosevelt as a forest management advisor (there were no national forests at that time), he invited Schenck to continue his work at Biltmore. Schenck accepted his invitation and quickly attracted a following of fourteen young apprentices, some of whom were relatively uneducated lumbermen and some college graduates.

As a result of his experiences with the young men, Schenck established a forestry school, for which he had to write his own texts because none were available at that time. Winter classes were held at the Biltmore Estate and summer classes at the forest. Classes at the forest were held in the community church building (Fig. 1–1). Mornings were spent in the classroom and afternoons in the field, where the students learned at first hand improved techniques for lumbering, soil conservation, and reforestation.

This important beginning of scientific forestry in America was made possible by George W. Vanderbilt, who acquired the forest land as part of his estate (one block of over 100,000 acres in 1895) and brought Pinchot and Schenck to manage it. At that time the Biltmore project (Pisgah Forest) was dubbed "Vanderbilt's Folly" by local citizens.

Among the many interesting discoveries of the present century is **photoperiodism**, the synchronous response of organisms to the length of their exposure to light. E. A. Schafer (1907) was the first to present evidence that the timing of bird migration was directly

FIGURE 1–1 *Replica of first forestry school in America. The school was established in 1892 by Carl A. Schenck on the Biltmore Estate, a part of which is now in the Pisgah National Forest, North Carolina.*

related to the length of the day. At the U.S. Department of Agriculture experiment station in Beltsville, Maryland, W. W. Garner and H. A. Allard discovered the same phenomenon in plants while crossbreeding varieties of tobacco from different parts of the world. They learned that the flowering of plants was triggered by the length of exposure to light and darkness and brought it to public attention in 1920. Incidentally, florists put this knowledge to practical use in forcing plants to bloom out of season.

The beginings of ecology consisted primarily of observations and explorations. Gradually ecologists shifted their attention from these largely qualitative endeavors to measuring, analyzing, and expressing their findings quantitatively. It naturally followed that once ecologists learned where organisms were found, they asked themselves why they were there. In the words of Paul Sears, "When the ecologist enters a forest or meadow, he sees not merely what is there, but what is happening there." In attempting to understand what is happening, ecologists have become increasingly involved in experimental research. Some problems that could not be investigated effectively with man's earlier state of knowledge can now be tackled. In recent years especially, some ecologists have focused attention on the tough problems of energy relations (energy flow through food chains).

Studies of community energetics have to take into account the amount of solar energy captured, biomass produced, input (import) from other communities, trophic levels and organisms comprising food chains, efficiency of energy transfers, and losses through respiration and removal (export). Much credit is due Edgar Transeau for his pioneering work (1926) on the photosynthetic efficiency of corn. Since then, others have measured the efficiency of entire communities such as old fields, lakes, and streams. The studies of Cedar Bog Lake, Minnesota, by Raymond Lindeman and of Silver Springs, Florida, by Howard T. Odum provided two models of special historical interest.

Unexpectedly, Rachel Carson sprung ecology from its cloistered atmosphere in 1962 with the publication of *Silent Spring*. What she had to say caught the interest of the public and started a wave of reaction against environmental pollution (Fig. 1–2). Perhaps she was actually articulating ideas that were already present in the minds of many people. At any rate, the publication of her book opened a new era of ecological concern and activity.

Still, the meaning of the word *ecology* was little known until the late 1960s, when people began to realize the extent of the environmental crisis due to a large and increasing population and the resultant pollution. On many college campuses students held "teach-

FIGURE 1–2 *Plants killed along right-of-way by herbicides. All the long-range effects of using biocides are unknown.*

ins" and demonstrated their anxiety about the deteriorating world environment. News media, public officials, and corporations joined the movement and showed their interest as never before. Thus far, there has been more verbal activity than action, but definite progress has been made. Americans are unlikely to lapse into a state of unconcern because conditions have become almost intolerable in some locations.

RELATIONSHIP TO OTHER SCIENCES

Ecology, like other branches of science, is not an endeavor all to itself. It deals primarily with biology, but other scientific disciplines are intermeshed with it. Naturally some are closer than others.

How different sciences combine to solve an ecological problem is illustrated by the "war" on pollution. Biologists may be measuring the sensitivity of organisms to contaminants while chemists are studying ways to detoxicate the poisonous substances. Physicists working on the same problem may be developing instruments to take rapid samples, whereas geologists may be searching for techniques to minimize the leaching of mine wastes into groundwater reservoirs. In addition, economists could be analyzing costs and lawyers might be settling liabilities. The point is that different talents combined into a team can often succeed where individual specialists cannot.

While ecology is primarily a biological science, several other sciences are primarily ecological. They cover a gamut of subjects from ancient climates to life in space. A few examples are **biogeography** (distribution of life), **paleoecology** (ecology of ancient times), **edaphology** (relationship of soils to plants), **exobiology** (life beyond the earth), and **physiological ecology** (physiological processes influenced by environment).

Because of the size of the field and the special interests of individual ecologists, the science of ecology has been divided in various ways. The specialties may be limited to dealing with times (**paleoecology**), places (**limnology**), types of organisms (**avian ecology**), objectives (**conservation ecology**), or levels of organization (**population ecology**).

SUBDIVISIONS BASED ON HABITAT

A convenient way to limit ecological study is by habitat. This approach emphasizes the environment as the determiner of the kinds of organisms present. As used here, habitat usually means a kind of habitat with similar conditions rather than a specific habitat restricted to one geographical location.

Arid lands ecology (deserts and semideserts) illustrates how extensive the unit may be. Although arid regions throughout the world support their own distinctive flora and fauna (Fig. 1–3), they have much in common. Life forms are similar because they have adapted to the same type of environment, and habitats are similar because aridity affects soils and atmospheric conditions in the same way. This means that the problems and opportunities for man in widely scattered arid zones are about the same.

A much more restricted branch of habitat ecology is **biospeleology** (cave ecology). A surprising number of animals and plants inhabit caves despite the darkness. Aside from their intrinsic characteristics, which are interesting to study, they offer a number of possibilities for utilization by man. For example, students at one college have used an old tunnel, which duplicates conditions in a natural cave, for ripening cheese.

These are many other branches of ecology based on habitat. Some of the larger and more familiar of these are **marine ecology**, **freshwater ecology**, **estuarine ecology**, **range ecology**, **grassland ecology**, and **terrestrial ecology**.

FIGURE 1–3 *A desert scene near Tucson, Arizona. The vegetation is characteristically scattered and consists of such plants as creosote bush, burro bush, ocotillo, palo verde, and succulents like saguaro and prickly pear.*

SUBDIVISIONS BASED ON TAXONOMIC AFFINITIES

Individual ecologists always have favorite organisms with which they like to work. Their personal preference may depend on where they have to go to find their subjects or under what conditions they can be investigated. As a result, some organisms have been studied more intensively than others because of their abundance or amenability to laboratory conditions. An "outdoor" person may find excitement in studying elusive mammals in out-of-the-way places, whereas an "indoor" person may prefer studying bacteria within the confines of a laboratory.

The broadest fields based on taxonomic relationships are **animal ecology** and **plant ecology**. These groups and their immediate subgroups **(insect ecology, avian ecology, bacterial ecology)** are much too broad for the individual researcher, however. The **ecology of pines, mosquitoes**, or **turtles** is more manageable. Better still, the latter might be subdivided into **ecology of marine turtles, fresh-**

water turtles, and **land turtles**, in which case the taxonomic group is studied by habitat.

SUBDIVISIONS BASED ON LEVEL OF ORGANIZATION

The wolf is a species threatened by extinction, partially because many people considered the animal undesirable and have offered a bounty for killing it. Fortunately, a few felt that the wolf's reputation as a sneaky, thieving beast was unjustified. They believed these animals had a role to play in the economy of nature and set out to find what it was. In one location studied, they found that wolves helped maintain a stable population of large herbivores. A study that concentrates on the relationship of the wolf or any other species to its environment is known as **autecology**.

The life histories of only a small portion of all the species in the world are known in any detail. Even if an autecologist makes a comprehensive study of a selected species like the wolf or the bald eagle, he must leave many questions unanswered. Any one inclined toward making autecological studies can find plenty to do.

A new dimension is added to the study of organisms when groups of individuals are considered collectively instead of individually. Whether the group is composed of a single species (population) or of mixed species (community, biome, or ecosystem), it is a functional unit having interrelationships among its own individuals as well as with surrounding abiotic factors. The ecology of groups of organisms is known as **synecology**. Because the scope of synecology is so broad, its subdivisions are often more useful as subject units. They are **population ecology**, **community ecology**, **biome ecology**, and **ecosystem ecology**.

Population is the name generally given to a group of individuals of the same species living together in a particular area, such as a population of fireweed in a clearing or selaginella on bare rock (Fig. 1–4). Being of the same "gene pool," the members breed freely among themselves. Their needs are virtually the same, forcing them into either a competitive or a cooperative interaction.

A number of everyday terms denote groups of closely related individuals that may or may not be the entire population of an area. Usually they are subunits of the population. In some cases, at least, the close relationship is beneficial to the individuals. Some familiar examples are colony (corals), pack (dogs), stand (trees), flock (birds), covey (birds), school (fish), pride (lions), bed (mollusks—e.g., oysters), den (snakes), nest (insects—e.g., hornets), and swarm (bees).

Another and broader way to look at the population concept is to

FIGURE 1–4 *Populations of plants. (a) A population of fireweed* (Epilobium angustifolium) *in Wisconsin; (b) A population (mat) of selaginella* (Selaginella tortipila) *in the southern Appalachian Mountains.*

consider a population as the entire membership or gene pool of a species. If some problems are genetically related, they might be handled better in the context of an entire species population.

A community consists of two or more (usually considerably more) kinds of organisms living together. The number of individuals or the size of the area is not important. What is important is that they share the same habitat and are interdependent. The community might be compared to a body and the component species to the organs of the body. Obviously, such a functional unit has fine-tuned interrelationships.

The community concept is an ancient one. While Theophrastus wrote about plant communities associated with specific environments, community studies have particularly flourished in this century. Community ecologists are studying such subjects as salt marsh communities, sand dune communities, old field communities, fir–spruce communities, and oyster bed communities (Fig. 1–5). Note that communities are named after habitats or types of organisms.

A **biome** is a composite of many interacting communities. Each biome occupies a large geographical area and is characterized by a distinctive form of life determined by the environment. Terrestrial biomes are somewhat better known than other types, especially by nonspecialists; therefore, this section deals with terrestrial biomes only. Biomes, though distinguished by vegetation types, include animals as well as plants. The various communities within a specific biome are unlike in different parts of the world; only their appearances are similar. Like the smaller communities, biomes are named after a type of habitat or organism (life form). If the latter, it will be the climax form (Chapter 3) for that particular climate.

A good way to learn more about biomes is to hold international conferences in which participants can share information. An unusually successful one, the International Conference on Arid Lands in a Changing World, was held at the University of Arizona, Tucson, in 1969. It brought together soil scientists, hydrologists, zoologists, botanists, conservationists, archeologists, geologists, geneticists, agronomists, sociologists, educators, etc.—all interested in deserts or semideserts. Unfortunately, the poorer nations were not well represented.

Another level of organization is the **ecosystem**, which is defined as a functional unit of biotic and abiotic components. To a degree it is self-contained—that is, it can exist with minimal interchange with other ecosystems. Examples are an aquarium, a lake, a river system and its contiguous watershed, or, in the largest sense, the earth itself.

FUNDAMENTAL ECOLOGICAL CONCEPTS

With the development of ecology as a science, there has been a steady evolution in what may be considered fundamental ecological concepts. It has become increasingly apparent that modern ecology has conceptual bases that are applicable to all practitioners in the field regardless of their special interests. These concepts are presented in a generalized form as follows.

FIGURE 1–5 *Natural communities. (a) A salt marsh community along the South Atlantic coastline. (b) An old field community dominated by broom-sedge* (Andropogon). *(c) A cypress community in Okefenokee Swamp. (Photo by L. P. Stratton.) (d) An oyster bed community.*

a

b

c

d

1. Structural and functional units. When both biotic and abiotic components are considered, the basic unit is the ecosystem. Discrete biological units consist of populations and communities (including biomes). Each population occupies a specific niche, a

unique functional position with respect to other organisms with which it interacts.

2. Interrelationships among organisms. Organisms exhibit varying degrees of positive, negative, or even neutral interactions with each other. These interrelationships occur at both inter- and intraspecific levels and determine, along with abiotic parameters, the degree of success a particular population has within a given habitat.
3. Energetics of ecosystems. Energy drives the ecosystem. This energy is captured by the autotrophic organisms (producers) and is transferred in the form of organic molecules to the heterotrophic organisms (consumers). This energy flow is unidirectional or noncyclic.
4. Biogeochemical cycling. The chemical components of the ecosphere move in defined cycles. Within the ecosphere, biological systems frequently regulate the rate of movement or cycling of the chemicals. Related to this is the role of water as the universal solvent for the biological systems.
5. Limiting factors. For success in growth and reproduction within a particular habitat, an organism requires various essential factors from its environment. The success of an organism is limited not only by deficiencies in substances or conditions but also by excesses. The minimum and maximum levels of tolerance for all environmental factors of a species vary seasonally, geographically, and according to the age of the population involved.
6. Succession. Ecosystems undergo an orderly process of change with time, passing from a less complex to a more complex state. This process involves not only changes in species composition but changes in the physical environment of a community as well. The terminal or stabilized state is known as the climax.
7. Disruption and exploitation. When natural conditions or acts of man reduce the species diversity of an ecosystem, the state of development is set back and the stability of the ecosystem reduced. Man's exploitation of ecosystems is directed toward channeling productivity to meet his needs.

An understanding of these ecological concepts grows with the study of the natural environment. The following chapters attempt to provide a basis for that understanding.

REFERENCES AND FURTHER READINGS

Allee, W. C., et al. 1949. *Principles of Animal Ecology*. Saunders, Philadelphia.

Boughey, A. S. 1971. *Fundamental Ecology*. Intex Educational Publishers, San Francisco.
Colinvaux, Paul. 1973. *Introduction to Ecology*. Wiley, New York.
Farb, P. 1963. *Ecology*. Time-Life, New York.
Kendeigh, S. C. 1961. *Animal Ecology*. Prentice-Hall, Englewood Cliffs, N.J.
Knight, C. B. 1965. *Basic Concepts of Ecology*. Macmillan, New York.
Kormondy, E. J. 1969. *Concepts of Ecology*. Prentice-Hall, Englewood Cliffs, N.J.
Kucera, C. L. 1973. *The Challenge of Ecology*. *Mosby*, St. Louis, Mo.
Odum, E. P. 1971. *Fundamentals of Ecology*. 3d ed. Saunders, Philadelphia.
Scientific American. 1971. "Man and the Ecosphere; Readings from Scientific American." Freeman, San Francisco.
Smith, R. L. 1966. *Ecology and Field Biology*. Harper & Row, New York.
Wagner, R. H. 1971. *Environment and Man*. Norton, New York.

Chapter 2 Development of General Awareness of Environmental Problems

In the past man has looked upon his environment as a limitless resource and has exploited it in the quickest, cheapest ways. One possible origin of this attitude can be seen in Genesis: God created man and said to him,

> *Be fruitful and multiply, and replenish the earth and subdue it. And have dominion over the fish of the sea, and over the fowl of the air, and over every living thing that moveth upon the earth. . . . Behold I have given you every herb bearing seed, which is upon the face of the earth, and every tree, in which is the fruit of a tree yielding seed; to you it shall be meat.*

A natural impulse of all creatures is to "multiply." (In fact, a criterion of ecological adaptation or "fitness" of a species is the success it has in reproducing itself in a particular habitat.) But only the human species is truly capable of having dominion over its environment and fellow species. Instead of viewing the good fortune of his favored position as the highest animal as a license for exploitation, man should interpret the injunction to "subdue" as a command to manage.

There is no doubt that as man increased in number, he had a considerable effect on his environment and its species composition. The hypothesis that man's "firedrive" method of hunting created the great grasslands and brought about the extermination of the "monster" mammals of the Pleistocene from much of the globe is strongly supported, if not proved. While early man appears to have had a role in the creation of the grasslands, his utilization of these areas was limited to hunting and perhaps the grazing of domesticated animals. The heavy sod of grasslands was difficult to cope with for agricultural purposes; thus, man's attempts to till the soil to grow his meager collection of domesticated plants was restricted to the flood plains of rivers. Here, the fertile soils were well watered and easily cultivated with stone or wooden tools.

After man had become nearly self-sufficient as a result of his agricultural efforts rather than relying solely on hunting, he could build towns and cities for his expanding numbers. With these concentrations of population, there came the need for intensified utilization of land for agricultural purposes, including grazing. Ultimately, dams and irrigation canals were built to meet these needs. It was in the valleys of the Tigris, Euphrates, Nile, and Indus rivers that some of the great early civilizations developed.

Six millennia of man's dominance along the banks of the lower Nile altered this area from the swampy African jungle that nature, without man, would have produced. Extensive irrigation systems were established in Egypt before 2000 B.C. The Aswan Dam, which has inundated some 5000 square miles, is only a modern continuation of a very long process—man's increasing ability to control his environment.

As would be expected, there were negative side-effects as early man began to congregate in cities. People in the crowded towns and cities were constantly subjected to epidemics and famines. Even at that time, pollution of the atmosphere and of lakes and streams was caused by smoke from man-made fires and erosion of burned and overgrazed slopes. Of greater significance is the fact that even then man was unable to anticipate the consequences of his own activities. The practice of using irrigation water to increase agricultural yields eventually led to the accumulation of high salt concentrations in the soil. As a result of declining soil fertility and the abandonment of land, silt was allowed gradually to clog the irrigation systems. Ultimately, the great civilizations supported by these systems vanished.

According to Pliny, the farmers of the great Babylonian Empire, which flourished before 2000 B.C., were able to harvest two grain crops annually and then graze sheep on the land between crops. Today it is fortunate for Iraq, part of the old empire, that there are abundant oil deposits within its boundaries, for less than 20 percent of the land is cultivated. The landscape is marked by the mounds of forgotten cities and old, silt-clogged irrigation works. The city of Ur, an ancient seaport, is now some 150 miles from the sea, and some of its buildings are buried under 35 feet of silt.

At some point in his cultural evolution, man learned that certain animals could be tamed and used to meet some of his basic needs such as food and clothing. It was much more efficient for him to herd domesticated cattle, sheep, or goats than to hunt the elusive wild game of the forest. However, most domesticated animals, differing from their wild predecessors, are not as well adapted to strenuous wandering in search of food and water. Thus, there is a marked tendency for domesticated herds to remain in a given area and to graze the range beyond the ability of the grass to recover.

The manifestations of overgrazed rangeland are seen today throughout the Middle East and along the Mediterranean Sea. The land once supported open forests with abundant grass on the mountain slopes and shrubs and grass on the plains. It is clear that even Plato was aware that deforestation and overgrazing could cause soil erosion and the drying up of springs. His warnings (in the *Atlanticus*) apparently were not heeded. Incessant grazing by ever-increasing herds, particularly of sheep and goats, eventually eliminated much of the forests by destroying the seedlings that would in time have replaced the old trees. Under increased grazing pressure not only the trees but even the grass disappeared. Without the trees and grass, the soil on the steep mountains washed down the slopes and settled in deep layers on the plains or heavily silted the rivers. The erosion was so extreme that mountain slopes in much of North Africa, Spain, Italy, Yugoslavia, Greece, and Turkey were stripped to bedrock. While the less eroded lower slopes of this region are frequently covered with a dense growth of evergreen shrubs, generally called garrigue or maquis, the present vegetation is of little economic value compared to the grass and forests it replaced.

It is not uncommon in the Middle East, the land of the cedars of Lebanon, to find the remains of ancient Roman roads standing several feet above the rocky desert. The road beds perched above the eroded countryside are grim reminders of the horror of erosion and the destruction of the forest of stately cedars for which the region once was famous. All but gone, even small stands of the cedars are hard to find today. In 1940 Dr. Walter Loudermilk of the U.S. Soil Erosion Service found a stand of the biblical cedars flourishing in an old churchyard where they had been protected from the grazing of goats for over 300 years.

In Central Europe the forests proved far more resilient than those of the Mediterranean. Rainfall there was abundant and evenly distributed throughout the year, resulting in dense, luxuriant forests of beech and oak. Widespread settlement of this region was delayed until man developed sufficient techniques for clearing the forests. This did not reach a peak until the height of the Roman Empire. After the dissolution of the Empire and during the prolonged period of civil and religious wars and plagues that followed, the forest once again advanced, forming a literal "dark age." Although the Renaissance again saw the clearing of the forests for farmland, the soil of this region remains fertile and productive even today. This would appear to be due to a tradition of long-term ownership and care of the land for centuries by a single family. Thus, the European forests made the transition from forest to field over a period of about 1000 years.

This was not the case for the North American deciduous forest. In the early seventeenth century the claim could be made that a squirrel with an inclination for travel could move from tree to tree from the Hudson River to the Mississippi without ever touching ground. Although there were clearings, due in part to the activity of the woodland Indians who burned areas in the forest to grow pumpkins, squash, and corn, this deciduous forest was extensive.

By the early 1800s much of New England had been cleared and consisted of crop- or pastureland, but not for long. Soon afterward New England farmers abandoned their poor farmlands and moved to virgin lands farther west. The abandonment of thousands of acres of poor, worn-out, and often badly eroded agricultural lands in the North and South alike became a common pattern in the settlement of North America. Fortunately, much of this area has been reforested. In recent years thousands of acres of land have been lost to strip mining. An additional million acres per year, according to conservative estimates, are covered by pavement, buildings, and other manifestations of man's "progress."

Both ecologists and laymen are frightened by what they see in the changing landscape and recognize that self-discipline in the use of natural environment must become a way of life. No longer can mankind defend the position that what nature provides is free for the taking, lest everyone suffer the consequences. For a long time nature has been successful in keeping the system balanced. Man's responsibility is to preserve that harmony as much as possible.

Ways to minimize disruptions within the natural landscape should and can be found by improving the methods by which man utilizes his environment. One who has observed the lumbering operations in a virgin forest will attest to this need. The noisy buzz of a chain saw followed by the crashing sound of falling giants of oaks and poplars is in sharp contrast to the beauty and serenity of the area a short time before. What was once enclosed in greenery is left open and scarred. Bulldozers deface the surface, falling trees crush the undergrowth, and discarded laps and tops clutter the ground. The scars from such an operation will last a lifetime. Better methods of harvesting trees must be found in order to conserve the vegetation and soil.

Second thoughts about the indiscriminate clearing of vegetation might be wise at this time, because no one knows how it may affect the environment. A look at the earth from an airplane shows that the vegetation has already been drastically altered (Fig. 2–1). Some land developers might do society a service by taking better advantage of surface contours and natural vegetation instead of leveling an entire area before starting a building project. Even the cross-country swath of a four-lane highway opens thirteen acres per mile (Fig. 2–2), disregarding land taken for additional right-of-way. In some areas the total land used for such highways is as much as forty-two acres per mile. Unrestrained clearing for whatever reason will surely bring on a crisis.

CULTURAL BASIS FOR CONFLICTS OF INTEREST

Although it is seldom articulated, the principal drive of mankind is the survival and maintenance of his species. In this respect he is like all other organisms. An intricate system of "checks" regulates the

FIGURE 2–1 *An airplane view showing how natural vegetation has been disrupted. (Photo courtesy of R. J. Holt, U.S. Department of Agriculture, Agricultural Stabilization and Conservation Service.)*

FIGURE 2–2 *A four-lane highway in a midwestern state. Such highways remove valuable land from productivity.*

influence one species can exert over populations of other organisms. They are kept under control by predator-prey relationships or other mechanisms. Man is an exception, however, in that his cultural evolution has allowed him to escape or circumvent some of the major biological restraints. In addition, man is the only animal capable of completely exterminating any other form of life.

Man has had to pay a price for the degree of cultural evolution he has attained. In order to accommodate his burgeoning population, he has had to develop techniques for providing food in the most efficient and economical way. This has meant the reduction of complex, natural ecosystems to much simpler systems, frequently to several species or even to monocultures.

Ecologists have long been aware that simplifying the environment to support only a few species can precipitate a catastrophe. Our highly mechanized agricultural operations, dominated by extensive acreages of a single crop, encourage the development of large and frequently persistent populations of insect pests or plant pathogens. A cornfield serves as a good example of the danger just described. To channel the potential productivity of a piece of land as directly as possible into a useful product (e.g., corn), man simplified the ecosystem by removing all plants or animals that might compete with the crop plants. In addition, he made the mistake of using extensively a genetic strain of corn that was susceptible to a fungal parasite. Consequently, in 1970 much of the crop was threatened by the southern corn leaf blight, race *T* of *Helminthosporium maydis* (Fig. 2–3).

FIGURE 2–3 *A corn plant damaged by the corn leaf blight* (Helminthosporium maydis). *The disease first reached epidemic proportions over much of the country in 1970. (S.C. Agricultural Experiment Station photo, courtesy of J. E. Halpin.)*

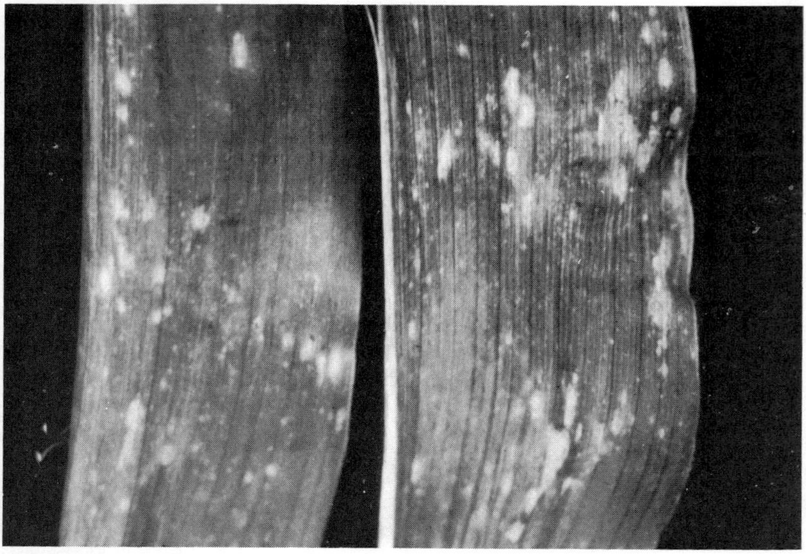

As insurance against insect damage, vast quantities of insecticides are applied with little regard for any other effects they may have, a practice now referred to as the shotgun approach. It should be noted that before World War II most of the insecticides used were nonpersistent organics found in natural environments. For example, the pyrethrins were derived from dried chrysanthemum flowers, nicotine sulfate from tobacco, and rotenone from tropical derris plants. Subsequently, with modern research, man has produced a number of potent, persistent chlorinated hydrocarbons (DDT, dieldrin, endrin, lindane, chlordane, heptachlor, and others) in his attempt to control the ever-present hordes of "harmful" insects, now considered to total some 3000 species. It has been estimated that there are over a billion pounds of DDT now circulating in the biosphere.

All of this effort to control insects is intended to increase the efficiency of man in utilizing his environment. The target organisms are indeed frequently controlled. That market yields and quality are increased by agricultural spraying is well documented. There is a growing body of evidence, however, that frequently repeated sprayings are merely insurance applications and in many cases unnecessary. Unfortunately, the farmer is being forced to this procedure in part by consumers who demand insect-free produce. This has now reached ridiculous proportions. Influenced by advertising, consumers now expect every fruit and vegetable they purchase to be a perfect specimen. They give no thought to the amount of environmental contamination that is necessary to attain such perfection. Perhaps it is in our common interest to relax our standards to a moderate degree, thereby permitting a great reduction in the quantity of pesticides used.

Another problem arising from man's cultural development centers in the growing affluence of an increasing portion of the population not only of the Western world but of the world in general. With greater purchasing power comes increased demand for goods and services. Frequently, the end result or by-product of increased affluence is waste from both production and consumption processes. The waste problem is complicated by the fact that our densely populated urban areas have high concentrations of heavy producers of substances that must be removed.

As a corollary to pollution associated with affluence, it has been said that while the 700 million relatively poor people of China might be bad for the country, the same number of relatively rich or affluent Chinese would cause severe environmental problems in no time. Although affluent people have the most time to be concerned about the quality of the environment, they are also the ones who occupy the most space, consume the most natural resources, bring about the most disturbances in the natural environment, and create the most pollution.

The challenge of the frontier has also played a significant role in shaping the attitudes of the American people toward their environment and provides another variable in analyzing the cultural basis for our environmental crisis. As the American pioneer moved westward, he faced many obstacles that he had to overcome. Many natural resources that are now highly valued he considered as liabilities rather than assets. For the farmer the forests had to be cleared, the marshes drained, and the rivers controlled. Too often, wildlife was considered a competitive threat to man. Without a doubt, the basic factor in this attitude toward the environment was the feeling that these resources were inexhaustible.

The first "conservation movement" developed around 1890—when the settlement of the frontier was nearly complete. The realization that our resources were indeed exhaustible brought about a new government philosophy toward natural-resource management. Theodore Roosevelt and Gifford Pinchot provided the leadership for this early movement.

A decided tendency in Western culture toward maximizing self-interest has also contributed to the impending crisis. While most people give lip service to the moral guideline "do unto others as you would have them do unto you," few have practiced it. For example, how can one municipality, without a qualm, dump its effluent wastes into the water of a river that another city downstream must purify at considerable expenditure of tax monies to make it usable? Until now man has applied the maxim *dilution is the solution to pollution,* without regard for the consequences. The volume of wastes has simply been too great for the natural system to absorb and purify, at least locally.

The decisions made by city engineers, corporation boards, state and national legislative bodies, and millions of private landowners ultimately determine the quality of the environment or the extent of its degradation. This, of course, is democracy exemplified. The responsibility is great, whatever they decide to do.

Another attribute of the current situation is the phenomenon of *institutional inadequacies.* The axiom that "what belongs to everyone belongs to no one" is particularly apropos to our current environmental problems. Because so many of our environmental resources, like air and water, appear to be plentiful, they are considered free commodities. Until rather recently, this country has been exceedingly permissive with respect to the use and abuse of these resources. There are several reasons for this. First, compared to other resources air and water are generally considered to have low economic value. Only under conditions of extreme impurity or scarcity do governing bodies place controls on their use and quality. Second, because these resources are not associated with a fixed location, right to use is more difficult to establish. Rights to use become delineated only when the resource is limited. Unlike most freshwater lakes and streams, the sky and the oceans belong to nobody.

Institutions, including local, state, and national governments, have produced a number of legislative acts and court decisions regulating or affecting the use of natural resources. It should be noted, however, that the determination to enforce existing statutes and to develop comprehensive controls has in general been lacking. Such indecisiveness is not difficult to understand. It is not simply a conflict between a few powerful special interests and "the people"; in fact, it is an issue involving a very large portion of the total citizenry implicated directly or indirectly in environmental degradation. The extent of responsibility ranges from that of the members of the board and executives of a chemical firm who condone the discharge of chemical waste into a river to the average citizen who opposes a tax increase for improving the municipal sanitation system. Thus, there develops a continuum of irresponsibility among institutions and individuals ranging from those directly responsible (e.g., some corporate executives), to the highly indifferent (e.g., some voters). With such a broad base of self-interests that feel threatened by more restrictive resource policies, it is not surprising that governmental units have been slow to meet the challenge.

Still another characteristic of man in Western society is his abiding faith in technology. After having sent astronauts to the moon, Western man tends to feel that technology can overcome almost any problem. He believes it is simply a matter of allocating sufficient quantities of money and brainpower. This optimism continues to exist in spite of substantial evidence that technology can and has produced harmful results, particularly in the long run. While there is no reason to depreciate the value of technology, it is an entirely different matter to view it as an environmental savior—an irrepressible force that will save man in spite of himself.

PUBLIC CONTROVERSY AND AWARENESS

In recent years, ecological catastrophes of varying degrees of seriousness have served as warning signals to the alert citizenry and to members of the scientific community. Efforts have been made first to collect sufficient data relevant to the particular environmental problem and then to determine the relationship between "cause" and "effect." In certain cases it is also a matter of establishing priorities, often between practices involving short-term convenience and long-term consequences. Because of the complexity of the environment, the relationship between cause and effect is often very difficult to establish.

In the past the public has looked to the universities and to various state and national agencies for advice and guidance concerning the use and consequences of insecticides, herbicides, wastes, etc., and are frequently dismayed by the degree of divergence in professional opinions. This is not surprising, because the nature of environmental problems is often very complex and intersects normal disciplinary lines of professional training.

Consider, for example, the "case of the Bolivian cats," which illustrates how complex a problem can be. The story begins with a severe epidemic of hemorrhagic fever (referred to as black typhus) in the upland Bolivian village of San Joaquin in 1963. During the course of this epidemic, which claimed the lives of more than 300 residents, the viral agent was identified. In the search for the vector of the disease, various organisms, including mosquitoes, mites, and ticks, were examined and eventually eliminated as carriers. In the search for other possible vectors, it was learned that there had been an unusual invasion of mouselike rodents caled *lauchas* (*Calomys callosus*) shortly before the epidemic broke out. These rodents had seldom been seen in San Joaquin before that time. Upon examination, the *lauchas* were shown to be the vector of the deadly virus. By trapping the rodents, the epidemiologists were able to control the spread of the epidemic.

The question, then, was: Why had these *lauchas* suddenly taken to living in human domiciles? Someone saw a relationship between the coming of the rodents and a declining cat population; until recently, cats had been ubiquitous in San Joaquin. However, at the time of the epidemic there were only about a dozen of them left among the 340 households in the village. The villagers indicated that their cat population had been on the decline for several years, victims of a strange "cat disease." The hypothesis that the cats had died of black typhus was disproved when in a subsequent experiment the cats were injected with the virus but did not catch the disease.

The real culprit (which indirectly caused black typhus) was eventually found to be DDT. For some time, DDT and dieldrin had been used in San Joaquin to control malaria. When the viral epidemic broke out, more DDT powder was applied throughout the village in order to kill the presumed insect vector. The cats apparently ingested DDT by licking and preening their fur, which had accidentally picked up the powder. The dosage was deadly. Before dying, the cats developed peculiar neurological symptoms that were consistent with DDT poisoning. Later this was confirmed by the U.S. Public Health Service. It was clear then that the DDT, which was brought to San Joaquin to control malaria-carrying mosquitoes, had brought about a completely unexpected result. The cat population was nearly eliminated, and, with their removal, the natural control of the wild *laucha* population was lost. For this lesson on ecological interactions some 300 villagers paid with their lives.

One need not look outside the United States to find other examples. In recent years American elms have been attacked by a fungus known as the Dutch elm disease, which plugs the water-conducting vessels of the tree and eventually causes its death (Fig. 2–4). A search revealed that two species of beetle transmitted the disease. Control, beginning in 1947, centered around the use of DDT. But while DDT killed the beetles that spread the fungus, it also killed desirable species.

In time observers noticed that bird populations in areas where

FIGURE 2–4 *American elms* (Ulmus americana) *killed by the Dutch elm disease* (Ceratostomela ulmi) *in Michigan (1970). The devastating disease is reported to be moving at the speed of twenty miles per year. The principal vector transmitting the disease in the United States is the European elm bark beetle* (Scolytus multistriatus).

DDT was commonly used were unfavorably affected. On the Michigan State University campus in East Lansing they found that the robin population had dropped from 370 to 4 birds in four years, and during this time few chicks were reared successfully by nesting birds. Bird populations elsewhere were found to be similarly affected. Subsequent study revealed that DDT was frequently being applied in doses much greater than necessary and that the soil under sprayed trees gradually accumulated DDT residues. Earthworms ingested the soil and thereby accumulated DDT in their bodies in concentrations exceeding that in the soil by a factor of ten or more. During the spring robins fed almost exclusively on earthworms and could have easily obtained a lethal concentration of DDT through this food chain.

When challenged, advocates of the use of DDT to control the beetle vector of the Dutch elm disease have simplified the issue by saying it is a choice between elms and robins. Yet the alternatives are not that simple. Spraying, even when properly done, can protect only a limited number of trees, and it is nearly certain that most elms will eventually succumb to this fungal disease. No single crusade against pests has raised passions so high or been waged with less justification than the Dutch elm disease program. As a result, the public is becoming increasingly aware of the DDT problem, and politicians are finding it expedient to be more responsive to the issue.

Now and then it is economics, rather than science, that revives a community's sense of values. For example, in the mid-1950s the land-locked salmon population at Maine's famous Lake Sebago diminished drastically. It was learned that the population decrease coincided with a considerable increase in DDT usage around the lake by owners of fishing camps and resorts who sought to protect fishermen and tourists from insect bites. While state wildlife biologists confirmed the significant decline in size and number of salmon and found a high incidence of DDT residues in the fish, it was the complaints of local merchants, who were losing money because of the poor fishing conditions, that brought about the abandonment of the use of DDT around the lake.

Public outcry has resulted in changes in several major programs sponsored by the federal government. The U.S. Department of Agriculture has all too frequently encouraged chemical control of insect pests rather than sound biological control practices. Although the USDA has insisted that those who apply pesticides follow "correct" procedures, large-scale efforts to "eliminate" several insect pests have caused great ecological disasters. These "correct" procedures presumedly were followed in widespread programs to eradicate the Japanese beetle and the fire ant (Fig. 2–5), neither of which has been successful to date. In response to the demands of a concerned public, the USDA has modified these programs.

The fire ant is a minor insect pest that reached the United States from Argentina in the 1920s. It gradually spread throughout most of the southern states without causing undue alarm. At worst, it was considered a nuisance. Nevertheless, in 1957 the USDA launched a massive propaganda campaign utilizing press releases, newspaper articles, and motion pictures to rally public support for programs to eradicate the fire ant. It followed up its verbal barrage with a deadly broadside of insecticides consisting of dieldrin and heptachlor, two of the most potent of the long-lasting chlorinated hydrocarbons. Initially the USDA sprayed the fire ant—and anything else that got in the way—with two pounds of heptachlor per acre over large areas of the Southeast. When criticized for this practice, the department reduced the dosage to 1¼ pounds per acre and eventually to ¼ pound per acre at intervals of three to six months. Finally, the USDA replaced heptachlor with mirex, which is less toxic and less hazardous to the environment.

FIGURE 2–5 *Harmful insects introduced into the United States. (a) Japanese beetle (Popillia japonica) on crape myrtle. The Japanese beetle was first observed in this country in 1916 near Riverton, New Jersey, and has widely extended its range since that time. Both larvae and adults are highly destructive to vegetation. They feed on many kinds of plants but are unusually destructive to roses, blackberries, and grapes. (b) A fire ant* (Solenopsis saevissima richteri) *mound. The fire ant was first found in Alabama in 1929 and has since substantially extended its range. (Photo courtesy of S. B. Hays.)*

a

b

Likewise, the USDA's massive program for the eradication of the Japanese beetle has been modified. Fortunately, biological controls have been effective in stabilizing the beetle populations. One such biological control involves a bacterium that attacks the beetle grub.

The massive pest control programs of the immediate past have come under closer scrutiny. Now major programs of eradication in which federal agencies take part must be examined first by the Federal Committee on Pest Control. This does not guarantee that other ill-conceived pest control programs will not be perpetuated, but the machinery has been established for careful advanced planning and justification.

As previously indicated, the relationship between an observed ecological problem and the possible causal factor(s) often cannot be readily determined. Consider, for example, the massive fish kills that were becoming an annual event in the lower Mississippi River and its tributary, the Atchafalaya, during the years 1958 through 1963. In the fall of 1960 some four million fish died. The principal victims were catfish, a primary source of food for many of the poorer people in the river towns and bayou settlements. The catfish were found bleeding around their mouths and fins and had enlarged swim bladders and digestive tracts. Although plenty of food was available for the fish, their digestive tracts were empty except for gas. Accordingly, the cause of death was labeled "abdominal dropsy." Water analyses failed to show any harmful substances or to give any clues to the cause of the deaths.

In the fall of 1963 another major fish kill occurred. This time various species of fish were affected, including fish-eating ducks such as mergansers. The magnitude of this kill was enough to halt commercial fishing. Because several million people along the Mississippi obtain their drinking water from the river, scientists from the Public Health Service were called in to investigate. Weeks of laborious analyses were carried out in an attempt to determine the nature and source of the toxic substance(s).

Early in the investigation it was ascertained that disease or parasites could be eliminated as the cause of the kill. Although laboratory mice fed tissues from the fish displayed symptoms of toxicity, the causes were not clear. But when healthy fish kept under laboratory conditions were fed residues from chloroform extracts of the river bottom sediments, they developed the symptoms of the dying fish of the Mississippi and Atchafalaya rivers.

By utilizing gas chromatography, it was found that extracts of both the fish tissues and the river mud contained traces of DDT and dieldrin, but not in significant quantities. Three other chemicals were present, which after much research were found to be endrin and two other compounds associated with the manufacture of endrin. One must remember that the polluted Mississippi contains literally thousands of organic compounds and that doses of toxic compounds such as endrin may be fatal to one species without having any visible effect on others. Eventually it was established that the endrin and its two related compounds were related to the selective killing of the Mississippi fish.

With endrin established as the primary culprit, the difficult task of locating the source began. Nearly six months after the fish kill, a team of pesticide experts from the Public Health Service followed the trail of endrin some 500 miles up the river to Memphis, Tennessee, where they found a firm that manufactured it. Water samples from sewers serving the firm were found to contain high concentrations of the substance. The firm denied any wrongdoing, and at that time the state and federal governments were not properly organized to meet the challenge of such an environmental problem complicated by scientific and legal ramifications.

As a result of the widespread attention given to the fish kill in the press and in hearings conducted by a U.S. Senate subcommittee, changes were made. Government investigators monitoring the pesticide content of the effluent of the Memphis plant found that the concentration declined gradually and eventually was no longer detectable. After considerable controversy, both among the public and within governmental agencies, changes in procedures were made so that future ecological problems might be more effectively handled. Since then, several ecologically concerned organizations have been formed that are prepared to engage in legal action to halt the activities of environmental offenders.

A good example is the legal action taken to protect the environment by restricting the use of the insecticide DDT. In the fall of 1968 two organizations, the Citizens Natural Resources Association, Inc., and the Izaak Walton League of America, Inc., requested a declaratory ruling in Wisconsin on whether DDT was an environmental pollutant. After extensive hearings lasting nearly six months, it was ruled that:

> *DDT and its analogs are therefore environmental pollutants within the definitions of Section 144.01 (11) and 144.30 (9), Wisconsin Statutes, by contaminating and rendering unclean and impure the air, land, and waters of the state and making the same injurious to public health and deleterious to fish, bird, and animal life.*
>
> *–Henkin, Merta, and Staples*
> *The Environment, the Establishment and the Law*

REFERENCES AND FURTHER READINGS

Benarde, M. A. 1970. *Our Precarious Habitat*. Norton, New York.

Boughey, A. S. 1971. *Man and the Environment: An Introduction to Human Ecology and Evolution*. Macmillan, New York.

Carson, Rachel. 1962. *Silent Spring*. Houghton Mifflin, Boston.

Ehrlich, P. R., and Anne H. Ehrlich. 1970. *Population, Resources, Environment; Issues in Human Ecology*. Freeman, San Francisco.

Falk, R. A. 1971. *This Endangered Planet*. Random House, New York.

Graham, F., Jr. 1966. *Disaster by Default: Politics and Water Pollution*. M. Evans, New York.

Graham, F., Jr. 1970. *Since Silent Spring*. Houghton Mifflin, Boston.

Henkin, H., M. Merta, and J. Staples. 1971. *The Environment, the Establishment and the Law*. Houghton Mifflin, Boston.

Hinrich, N. 1971. *Population, Environment, and People*. McGraw-Hill, New York.

Love, G. A., and Rhoda M. Love, eds. 1970. *Ecological Crisis*. Harcourt Brace Jovanovich, New York.

Nicholson, M. 1970. *The Environmental Revolution*. McGraw-Hill, New York.

Shepard, P. 1967. *Man in the Landscape*. Random House, New York.

Part 2
Natural Communities, Habitats, and Man

Chapter 3
Forests

RELATIONSHIPS OF LIFE FORMS TO THEIR ENVIRONMENTS

In distinguishing one kind of organism from another, the emphasis is on its individuality—how it is unlike other organisms. Its uniqueness and status among associated organisms form the basis for the concept of the **niche**, the position filled by a population of organisms (species) in a community. Yet most species are surprisingly similar when it comes to coping with the same set of environmental circumstances for the simple reason that adaptation means survival and failure to adapt means extinction. The survivors, selected by common environmental stresses, are called **life forms**, a term used for characteristic groups of organisms such as grasses, deciduous trees, conifers, and so on.

That every life form is closely related to the environment is obvious. The environment includes climate, soil, and other organisms. Chief among these is climate, which plays a major role in determining even the kinds of soils and organisms. Climate itself has many aspects, the most important being temperature and water. These two together, in quantity and annual distribution, set limits by providing conditions in which a particular life form can or cannot exist.

Terrestrial plants are generally accorded more emphasis than the animals associated with them because they are more fundamental in the community structure and easier to sample. They are at the base of the food pyramid (see Chapter 7) and furnish shelter for most animals. Just as plants are related to climate, animals are highly keyed to specific kinds of plants. For examples of close animal–plant relationships, consider grazing animals and grass or arboreal animals and trees.

THE BIOME

A community is a biome when it has a regional dimension and its dominant members are characterized by a specific life form. Since it occupies a climatic region, it is the largest clearly identifiable community. In this sense its meaning is identical with that of **formation**. Some ecologists have expanded the meaning to include aquatic communities (e.g., coral reefs) as well. Terrestrial biomes, though characterized by plant life forms, also include animals. At present the International Biological Program (IBP) has major studies under way in six such biomes: grassland, deciduous forest, tropical forest, coniferous forest, tundra, and desert.

The term *biome* may be used in an abstract or concrete sense just as the distinction can be applied to smaller communities. Thus, a rainforest may be a specific one like the Amazon rainforest (concrete) or the mental image of one synthesized from a knowledge of all rainforests (abstract). In the latter case it is a composite from different geographical areas.

STABLE AND UNSTABLE COMMUNITIES WITHIN THE BIOME

A biome is composed of both stable and unstable communities. The stable ones, called **climax communities**, display the typical life form, whereas some of the unstable ones may not. A climax community is delicately adjusted to some predominant long-term natural cause, especially climate, and reproduces itself indefinitely. A climax forest contains trees of all ages and sizes from seedlings to old trees. As the old trees die, younger individuals of the same species are always available to replace them.

Unlike the stable climax community, unstable communities do not reproduce themselves, at least for long. In these transient communities, organisms invade suitable places and remain there until conditions change. Even if the climate remains the same, conditions change because of weathering and erosion processes as well as alterations by the community itself. As these factors modify the environment, it becomes less favorable to the existing community and more favorable to another. Eventually a new community replaces the earlier one. The replacement of some communities of organisms by others is called **succession**. It begins with **pioneers** that enter unoccupied territory and become established, and ends with a stable climax community.

Disturbances of primary climaxes are so commonplace now that modified communities and successional stages are replacing them at a rapid pace. A climax altered by the destruction of endemic species that were formerly dominant or by the replacement of dominants with immigrants is a **disclimax**. Like the true climax, it is a stable community, but it is held that way by continuous or recurrent disturbances. The likelihood is that most disclimaxes would proceed to their original status if the disturbance were removed. Man and domesticated animals are primarily responsible for the disturbances causing a disclimax. Such disturbances are caused by cultivating, lumbering, burning, grazing, and introducing new organisms.

A disclimax is sometimes called a **subclimax** because it might be considered a prolonged successional stage preceding the climax. The cause of a subclimax does not necessarily have to be a disturbance, however; it may be due to prolonged development. In general, the term *subclimax* is used to refer to any stage preceding the climax.

Outline of Major Forest Biomes

- Low-latitude forests
  - Tropical rainforest (selva or great forest; also includes gallery forest and jungle)
  - Mangrove forest
  - Semideciduous forest (monsoon forest)
  - Scrub forest (thorn forest or bush country)
- Middle- and high-latitude forests
  - Temperate rainforest
  - Mediterranean-type scrub forest (chaparral, maquis, garrigue, or broad-sclerophyll vegetations)
  - Deciduous forest
  - Northern coniferous forest (boreal forest or taiga)

TROPICAL RAINFOREST (SELVA OR GREAT FOREST)

Nowhere on earth are conditions more favorable for plant growth than in the equatorial region, where both temperature and rainfall are high. The tropics also have the distinction of having more species than any other region; the number dwindles toward the poles. Variety of life, uniformity of environment, and optimum growth conditions make the rainforest lush and spectacular (Fig. 3–1).

In area, the rainforest covers more than 10 percent of the land

FIGURE 3–1 *Tropical rainforest in Bolivia. (Photo by S. T. McDaniel.)*

surface of the earth and nearly 50 percent of its forested land. Three large areas, the Amazon, the Congo, and the East Indies, comprise the bulk of the forest. Smaller forests are in Central America, the West Indies, Indonesia, and some oceanic islands. Tropical areas not blanketed by rainforest have topographic irregularities and atypical wind or oceanic currents, conditions that are unfavorable to rainforests.

At the equator, the air of the converging easterlies becomes lighter as it expands under the influence of higher temperatures and rises. Upon reaching high altitudes the air cools, and moisture begins condensing to rain. The annual rainfall is about 90 (80–120) inches but is known to have exceeded 450 inches. Its distribution throughout the year varies from uniformity to distinctly wet and dry seasons. During the driest month of the so-called dry season, the rainfall is seldom less than four inches. For many consecutive days torrential rains may drench the forest. For the most part, the winds are easterlies that have traversed large bodies of water and picked up large quantities of moisture.

Contrary to popular belief, the temperature of rainforests is not unduly high. At a station in Central Brazil, the highest temperature recorded was 98.4°F, the lowest 69.4°, and the average 81.0°. Someone has estimated that the mean annual temperature of rainforests is around 78°.

High humidity and perpetual summer warmth favor rapid growth and decay. Leaves drop one by one and quickly decompose, leaving little or no accumulation of humus. Without this cover of humus, bare clayey soil is a common sight. The soils are best described as impoverished. They are so acid that necessary elements such as calcium and potassium are dissolved and lost. The heavy rains leach the soluble materials from the surface soil, leaving it rich in iron and aluminum. Such a soil is a **laterite**.

Woody trees draped with vines (lianas) give the forest its distinctive feature. They seem to be stretching and scrambling skyward, grasping for light. Some vines, like the strangler figs, start at the tops of trees as epiphytes (plants which grow on other plants) and send roots downward until they make connections with the soil. The latticelike roots cling to the trunk of its host, soon encasing it (Fig. 3–2). Once connection is made with the ground, the stranglers grow much faster and eventually kill their host. Many smaller plants that are unable to grow so tall solve the problem of obtaining sufficient light by perching on trunks and branches high up in the trees, making the forest a veritable hanging garden adorned with such striking plants as orchids, bromeliads, and ferns (all present in the Western Hemisphere). These epiphytes have waxy leaves that retard evaporation. They also have highly absorbent spongy roots that drape over the limbs and water reservoirs at the bases of leaves that serve them well in capturing and retaining water. Their flowers tend to be brilliantly colored. They are, however, difficult to see and enjoy at their great height above the ground.

Contrary to popular opinion, the forest floor is left uncluttered by undergrowth simply because the twilight intensity of light is insufficient for adequate rates of photosynthesis. The dense canopy above entraps the heat of day by interfering with reradiation and retards evaporation by sheltering the lower forest from wind and sun; therefore, the air between floor and canopy is muggy.

The forest profile consists of three or more strata (Fig. 3–3). The trees of the **canopy** vary from 60 to 120 feet in height, and life is concentrated in this stratum. Beneath the canopy are one or more strata that may be lumped together as the **understory**. And towering over the canopy are the scattered **emergents**, which sometimes reach a height of 200 feet or more. Observing this stratification, Humboldt described the rainforest as "a forest above a forest."

Rainforest trees are broad-leaved evergreens. The trunks of large trees are characteristically buttressed at the base or have above-ground roots that radiate like buttresses. Buttresses up to twenty feet high give needed additional support for tall trees with shallow root systems. It happens that rainforest trees are generally shallow rooted, because the typical rainforest grows in lowlands (below 1000 feet in

FIGURE 3–2 *A strangler fig on a palm tree. Beginning as an epiphyte, the fig grows roots to the ground. The roots encase the trunk of the host tree and may eventually kill it.*

FIGURE 3–3 *A rainforest profile. The trees occur in three or more strata. The uppermost continuous stratum, the canopy, is overtopped by scattered giants called emergents.*

elevation), where the water table is near the ground surface. In such waterlogged soils oxygen is a limiting factor for the development of deep roots. Another adaptation for anchoring trees in the soggy soil is stilt roots. Trees in the lower strata usually have leaves with drip-tips (Fig. 3–4) that facilitate runoff during heavy rainfall.

It is probably safe to say that for any given area in a tropical rainforest the number of tree species would be at least ten times as great as in similar-sized areas in forests in the temperate zone. A single square mile may contain as many as 3000 species of trees and shrubs. Unlike the situation in many other types of forests, the trees are not in pure stands. Instead, members of the numerous species are individually distributed, a pattern that makes lumbering more difficult and expensive. Some of the wood is unusually dense and hard. A few of the better-known rainforest trees are ebony, mahogany, screw pine, rubber, rattan palm, camphor, and cacao (source of chocolate).

Animal life is abundant and highly adapted to the many types of habitat offered by the large variety of plants, the level of canopies, and epiphytic microhabitats. Predominantly arboreal, animals tend

FIGURE 3–4 *A rainforest leaf with a drip-tip. Water drains away at the tip like water from a spout.*

to be either brightly colored or well camouflaged. Many of them have grasping paws, prehensile tails, and adhesive pads with which to cling to trees; others have skin folds that can be used as wings to glide from place to place. Many of the New World monkeys have prehensile tails, an innovation not shared by monkeys of the Old World. A smaller number of rainforest animals are terrestrial. By day the animals do not seem as abundant as they really are, their presence often concealed and their noises subdued; but at dusk and dawn the forest seems to awaken with activity as nocturnal and diurnal animals mingle as if they were changing shifts.

In habitats unsurpassed in number or complexity, insects and spiders find their own little worlds. Of all the groups, insects are the most common. Among them are many kinds of ants, termites, beetles, butterflies, and moths. Many are much larger than their counterparts found in temperate regions. For one thing, their wide wingspan is advantageous in the stillness of the forest but impossible in more open places where heavy winds would tear their fragile wings to pieces. The development of mimicry is often spectacular, as in the case of a katydid that looks like a leaf adorned with dew-drops and fungal spots.

Representative vertebrate animals in the American rainforest are toads, lizards, snakes, turtles, birds, monkeys, tapirs, peccaries, anteaters, opossums, armadillos, otters, raccoons, deer, squirrels, sloths, coatis (raccoonlike), porcupines, ocelots, pumas, jaguars, and bats, not to mention the fishes.

The birds, in particular, run the gamut of variation in habits, size, and color from the delicate hummingbird to the large vulture. Among the more familiar are parrots (including parakeets and macaws), woodpeckers, toucans, owls, eagles, and hawks.

Distinctive rainforest animals are monkeys, apes (chimpanzees and gorillas), lemurs, hogs, rodents (porcupines, squirrels, and many others), birds, chameleons, and representatives of most other groups. Noteworthy insects are termites, bees, ants, and brilliantly colored butterflies.

A few animals (mammals) found in the rainforests of Southeastern Asia are leopards, flying lemurs, bats, monkeys, tree shrews, and mouse deer. The gliding adaptation to the arboreal way of life is found in squirrels, snakes, frogs, lizards, flying lemurs (colugos), and bats—more gliding animals than in any other region. The Southeast Asia rainforest is estimated to contain more than 1500 identified vertebrate species, including 213 species of snakes, 148 lizards, 182 frogs, and 351 fishes.

Except for a few aborigines, like the Pygmies in the Eastern Congo, man has considered the rainforest inhospitable. As time passes, he exploits it more and more for lumber and other natural products or clears it to grow rubber, coffee, cacao, bananas, sapodilla, and the like.

Opening up the forest lets in the light, which allows undergrowth to flourish. Sometimes extensive damage to the forest results in its replacement by scrub forest or savanna. Once disturbed, restoration of the forest to its original state is a very slow process.

The rainforest we have described is a typical lowland forest. It may, however, take many other forms as a result of topographical irregularities, drainage patterns, soil differences, tides, and disturbances. Some of the various types are called swamp rainforests, dry rainforests (because they have better drainage), montane rainforests, gallery rainforests, mangrove forests, and jungle. All may be considered different aspects of the rainforest either because they grow under the same climatic conditions or because they contain broad-leaved evergreen life forms. A look at three of the variant types will suffice to show how environmental differences can modify the forest in distinctive ways.

The **jungle**, through which man must hack a trail as he penetrates it, is probably the image most people have of the rainforest. In part this may be because the average traveler, having no business in the forest, sees it from the fringes, where it really is a jungle. Should he take the time and trouble to go beyond the jungle fringe, he would find the space among the towering tree trunks unobstructed. A jungle is simply a rainforest densely cluttered with undergrowth.

Jungle is caused by one thing: sunlight getting through or under the canopy to the forest floor. There are three conditions in which this is possible. First, the edges of clearings, streams, or bodies of water result in breaks in the canopy, allowing sunlight to penetrate from the sides. Consequently, a large river like the Amazon will be bordered by jungle. The same is true around villages, fields, or natural openings like those caused by wind-throws or the death of large trees. Second, destruction of a rainforest is often followed by replacement with secondary growth (jungle). Third, jungle develops on steep slopes where crowns of trees are not interlocked in a continuous canopy. Since on a slope the crowns of trees have about the same magnitude of vertical displacement as their bases, light can penetrate from the side and support undergrowth.

The **gallery forest** (galeria) is a stream-bank "rainforest" that often continues upstream into other forest types or grassland. Like ribbons of green, these forests penetrate more arid territory where the rainfall is insufficient to support a rainforest. The rainforest life form is sustained by water in the streams, like the fringe of cottonwoods or salt cedars along some streams in the arid West or brush and trees beside an irrigation ditch. Where the stream is wide, the two strips of forest on each bank are separated so that sunlight penetrates well and results in the formation of jungle. Further inland, where streams are smaller, tree crowns from each bank merge overhead, enclosing the streams in tunnels of greenery that give the forest its name. Owing to the effect of shading, undergrowth is minimal along the stream passage through the gallery.

MANGROVE FOREST

Throughout the tropics where the rainforest meets the sea a fringe zone that may be ten or more miles wide is alternately flooded at high tide and exposed at low tide. Within this zone is the **mangrove forest**, an entanglement of broad-leaved evergreens commonly perched on stilt roots. The trees are not as large as in the typical rainforest. Although some may reach a height of 100 feet, many are shrublike. The mangrove forest acts as a buffer against stormy winds and waters, protecting the inland forest behind it. At times, stretches of the forest are killed by severe storms. Because of the dense, tangled vegetation and quaggy bottom on which it stands, mangrove is perhaps the most difficult of all forest types for man to penetrate and utilize.

Mangrove is not the name of a specific tree—it is the name given to various species with the same life form. In south Florida there are three mangrove species—red, black, and white—all belonging to different taxonomic groups. In Asia there are about thirty species of mangrove trees.

Mangroves grow best on a muddy substrate and poorest on sand. Their favorite habitats are coastal mud flats and marshes, where waterborne sediments settle as the rivers carrying them lose momentum. From the ocean the saltwater tides sweep inland, and from the land the freshwater drainage flows outward, subjecting the zone to wide fluctuations in salinity to which the plants are notably tolerant. The waterlogged soil often reeks with the rotten-egg odor (hydrogen sulfide) produced by the anaerobic respiration of bacteria. The soil is so deficient in oxygen that many of the trees have special adaptations to increase the supply.

Mangrove root systems are usually specialized in one of three ways—with stiltlike main roots (Fig. 3–5), with above-ground arches of horizontal roots, and with vertical stalklike branches projecting above the ground from subterranean horizontal roots. One such tree is the black mangrove of Florida, which sprouts many vertical branch roots called "breathing tubes" that resemble soda straws in size and shape (Fig. 3–5). Parts of all of these above-ground root systems, whatever their form, facilitate the exchange of gases that occurs in respiration.

Mangrove trees are effective soil binders in a zone of instability washed and pounded by waves, tidal flow, and tropical storms. They literally advance on the sea, reclaiming soil as they go. Reclamation and soil building results from the action of the mangrove in binding soil particles together, sheltering the soil from some erosive forces, and catching and holding debris. The close growth of interlocking spidery roots, so characteristic of many mangrove trees, and the vertical projections from some root systems are perfectly suited for entrapping water-transported drift without seriously impeding the back-and-forth movement of tidal water. At the same time, slowing the movement of the water causes the settling of silt and clay that hastens the soil-building process.

The red mangrove, also in Florida, has a unique way of reproducing. The seeds sprout while still in the pod and grow long hypocotyls (stalks) that project 8 to 12 inches from the bottom. They hang pendulously from the branches (Fig. 3–6) and plant themselves in an upright position if they fall in the mud. If the embryos fall into water, they float perpendicularly, increasing the possibility that

FIGURE 3–5 *Mangrove roots. (a) The stiltlike roots of a mangrove tree in the Florida Everglades. (b) Vertical branch roots of the black mangrove of Florida. The vertical roots facilitate the absorption of oxygen.*

a

b

FIGURE 3–6 *Reproduction of red mangrove. (a) A seed pod from which the hypocotyl of an embryonic plant projects. When the pod falls from the tree, the projecting hypocotyl usually points downward and often embeds itself in the mud in an upright position; thus, the embryo is naturally planted. (b) Mangrove seedlings naturally planted in the mud beneath a red mangrove tree.*

a

b

they may come to rest in a position most favorable for root growth. The young plants, like the mature trees, endure salt water well.

Growing with the dominant trees are a variety of salt-tolerant plants, especially grasses and succulents. Succulents found in the Everglades are sea-blite, pickleweed (*Suaeda*), glasswort (*Salicornia*), and sea-purslane (*Sesuvium*). The leather fern, the largest North American fern, also thrives in the same habitat.

Animals abound in number and variety. Birds are everywhere. Particularly well represented are birds that feed on fish and other aquatic life. These birds congregate in rookeries, where their droppings accumulate to such an extent that they often damage or kill the trees. The ground around the roots is a haven for insects, spiders, and innumerable crabs. When the tide is in, or in tidal pools and channels when the tide is out, quite a variety of fish may be found. The shallow waters of Florida mangrove forests are favorable nursery sites for shrimp, shellfish, and a number of economically important fishes like snapper, drum, mullet, snook, and menhaden. In similar places in Asia are two unique fishes—the mudskipper, which ventures on land and up mangrove plants, and the archer fish, which shoots its prey with a stream of water. Among the larger animals that venture into the mangrove forest are alligators and crocodiles.

SEMIDECIDUOUS FOREST (MONSOON FOREST)

The **semideciduous forest**, primarily tropical, grows where there are monsoons or in other places where a seasonal deficiency of water is a limiting factor. In total biomass it stands between the rainforest and the scrub forest. Some semideciduous forests are found in South America, Africa, Central America, and the West Indies, but the largest grow in Southeast Asia and surrounding areas where the monsoon climate prevails. For that reason the semideciduous forest is often called the **monsoon forest**, especially in Asia.

The monsoons are seasonal winds that reverse direction twice a year, alternately bringing rain and drought. In general, people primarily envision the coming of rains; hence, seasonal rains are also called monsoons. The character of this forest—alternately flourishing and withering—is a direct result of the climate. During the wet season rainfall may be torrential and seemingly endless, falling day after day and causing the ground to become soggy, land to slide down slopes, and streams to overflow their banks. The vegetation responds with vigor. About six months later, shifting winds reverse the seasons, bringing a parching drought during which the leaves turn brown as they gradually die; some of the trees become deciduous like eastern American trees in the winter.

The monsoons are most distinctive in the Southern Asiatic–Northern Australian sphere. Other monsoon areas are comparatively small. In the Asiatic–Australian sphere the landmasses and oceans are geographically disposed in such a way that strong winds are generated when the full impact of heat from the sun shifts from south to north across the equator as winter becomes summer. The large landmass north of India, for example, heats rapidly during the summer and produces a low-pressure area as the atmosphere expands. At the same time, across the equator to the south a high-pressure area develops over the Indian Ocean, where it is winter; remember, too, that air warms up more slowly over water than over land. The contrast between the high atmospheric pressure in the oceanic area below the equator and the low pressure in the landmass above it is so great that strong easterly trade winds come sweeping across the equator over the northern trade wind belt, contrary to the usual midglobal pattern. They come toward the equator as easterlies (from the southeast), but because of the rotational effects of the earth, they bend in the other direction after crossing the equator and enter India from the southwest.

Having traveled over the ocean, the monsoon winds are heavily laden with water. As they rise over irregular landmasses and are cooled, water is condensed to rain, causing about 85 percent of the annual precipitation to fall in one season. Although semideciduous forest is widespread in India, the monsoons alone do not determine the type of vegetation; other types of biomes are present. In addition to seasonal distribution of rain, the amount of rainfall, soil conditions, topography, and evaporation produce different environments.

The arrival of winter reverses the pattern, causing winds to cross the equator from the Northern Hemisphere to the Southern. Then the monsoon winds bring heavy rains to some parts of the East Indies and northeastern Australia. At the same time dry continental winds reach India from farther north.

Trees in the semideciduous forest are broad-leaved like those in the rainforest, but most are partially or completely deciduous during the dry season. Many are rainforest species that can also tolerate the less favorable environment. They are smaller, wider spaced, and sometimes deciduous, and they often grow in pure stands. The more open canopy allows greater sunlight penetration and favors the growth of underbrush; therefore, much of the forest is jungle. The pure stands of timber attract lumbermen, who exploit them extensively. Teak in particular is taken from semideciduous forests, where its quality is far superior to teak from the rainforest.

Animal life is abundant. Some animals, like insects, birds, and reptiles, are actually more abundant than in the rainforest. Larger animals that forage on bordering grassland retreat to the forest for shelter. Numerous herbivores of all sizes come and go, followed by a host of carnivores that prey on them. Some of the herbivores are elephants (Africa and Asia), giraffes (Africa), zebras (Africa), wildebeests (Africa), oryxes (Africa), and impalas (Africa); the carnivores include lions (Africa and Asia), tigers (Asia), leopards (Africa and Asia), and cheetahs (Africa and Asia). The lumbering elephants, trampling the vegetation and browsing and grazing to satisfy voracious appetites (sixteen hours a day), clear pathways through the tangle of trees in forests bordering grasslands. This makes it easier for other animals to utilize the forested zone. Some places suited for semideciduous forest are densely populated by man. He has, of course, destroyed or drastically altered the original forest.

SCRUB FOREST (THORN FOREST OR BUSH COUNTRY)

The **scrub forest**, also known as bush country or thorn forest, occupies areas too dry for rainforest or semideciduous forest. Usually the annual rainfall is under forty inches. Were it a little drier or repeatedly burned, grassland would prevail instead of trees. Actually the forest varies from a thin stand interspersed with grass to a thick, almost impenetrable entanglement of trees, shrubs, and vines, the character of the stand depending on site and climatic conditions. In any case, the trees are much smaller than those growing in more favorable locations. At best the vegetation looks much like scrubby second growth.

The trees have adaptive characteristics that are related to the arid conditions under which they live. Their leaves are usually small, an adaptation that reduces loss of water. In the dry season the trees shed their leaves—another of nature's ways to check water loss by transpiration. They remain inactive until water becomes available again. The trunks of these trees are insulated with relatively thick cork. Because so many of its plants are thorny, this forest is often appropriately called thorn forest. Thorns are modifications of different morphological structures; the large paired thorns on some leguminous trees in Mexico are noteworthy. These are modified stipules (appendages found at the bases of some leaves) like those found on the black locust trees that grow in the United States from the Plains states to the Atlantic Coastal states. Since plants in arid regions are often armed with prickles, briers, and thorns, these structures must give them a survival advantage needed in a harsh environment.

The frequent juxtaposition or intergradation between scrub forest and savanna makes the forest a sheltered retreat for animals that come from the open country seeking shade and concealment. Many, like antelopes and zebras, have spots or stripes for camouflage among the shadows. At the approach of danger, such animals immediately freeze in position so that they become difficult to see. The margins between savannas and scrub forests in Africa have the greatest density of large herbivores of any place on earth. Sometimes there are as many as 100 large animals per square mile, representing essentially the same species as in the semideciduous forest. The climate is not ideal for man, although many have to endure it in overpopulated Asia. Even in Africa, land occupied by scrub forest is increasingly utilized for farming.

TEMPERATE RAINFOREST

The temperate rainforest is quite different from the tropical rainforest, although both receive abundant rain well distributed throughout the year—that is, without a dry period severe enough to force seasonal dormancy. Of the two seasons, winter is wetter than summer. Precipitation is mostly in the form of cold rain, although some falls as snow. Annual precipitation may go well above 150 inches. In addition, the leaves are often drenched with mist or fog for many consecutive days.

Geographical conditions causing the predominant weather pattern that supports the forest are west coast mountain ranges that convert incoming westerly winds into updrafts. The winds are heavily laden with moisture, which precipitates as the ascending air is cooled. Rising air currents on western mountain slopes cause the rainforest in the narrow coastal strip from northern California to Kodiak Island, Alaska. Other temperate rainforest locations are southern Chile, Tasmania, the west coast of New Zealand, and southeastern Australia.

The temperate rainforest is primarily evergreen and coniferous. Whatever the regional species, the towering trees normally form a closed canopy through which sunlight penetrates dimly. The trees are giants in height and girth. Some, like the redwoods, easily exceed 300 feet in height and 60 feet in circumference. Except where the canopy is interrupted, woody undergrowth is minimal. In the enclosed darkness, mosses and liverworts clothe the trees and deeply carpet the forest floor. Delicate, filmy epiphytic ferns do well in the humid environment.

Because of their great size and age, the redwoods at the southern end of the American forest are well known. Where they grow, the summer is dry except for daily fogs that sweep in from the Pacific and drench the leaves, providing ample moisture for growth and reducing transpiration.

In the cool temperature of middle-latitude rainforests, decay is retarded so that a considerable amount of organic matter, litter, and humus accumulates on the forest floor. Many trees are uprooted where the soil is soggy and shallow, which is often the case in the coastal mountains where much of the forest grows. The roots, spread horizontally near the surface, do not form a good anchorage for tall trees. Crisscrossed logs of toppled trees pile up (Fig. 3–7) and create obstacles for animals moving within them.

The North American rainforest is unusually spectacular because of its giant conifers. A good place to view it is Olympic National Park, on the Olympic Peninsula of Washington State, where there are dense stands of Douglas fir, western red cedar (also called giant arbor-vitae), western hemlock, and Sitka spruce. Some plants of less importance are the big-leaf maple (an understory tree with the conifers), vine maple, wood sorrel (*Oxalis*), and ferns.

FIGURE 3–7 *The crisscrossed trunks of fallen trees. This particular scene is a fir–spruce forest along the Appalachian Trail in the higher mountains near the Tennessee–North Carolina border. In a similar way, trees are toppled in other types of forests where root systems are near the surface because the soil is shallow or constantly wet.*

The Olympic Forest climax consists of western red cedar, western hemlock, and Sitka spruce. A surprising number of the hemlocks and spruces are on stiltlike roots. The reason for this odd habit is that seedlings can hardly penetrate the ground cover of mosses and other plants but can get started atop fallen logs. As they grow, their roots have to straddle the logs to reach the ground. In time the logs rot, leaving the trees perched above the ground.

The large stands of Douglas fir are subclimax, and their seedlings are unable to survive in the dense shade of large trees. Destructive fires, often following logging operations, provide the ideal opening and seedbed for the firs. Although the rainforest is usually humid, it can become quite dry after a drought of as little as two or three weeks, and then the dry moss and resinous conifers are easily kindled.

Large mammals of the Olympic rainforest include Roosevelt elk, black-tailed deer, black bear, and cougar.

MEDITERRANEAN-TYPE SCRUB FOREST (CHAPARRAL, MAQUIS, GARRIGUE, OR BROAD-SCLEROPHYLL VEGETATION)

In the temperate zone of both hemispheres are widely scattered areas where the vegetation consists predominantly of broad-leaved evergreen shrubs and trees. This is **Mediterranean-type scrub forest,** so named because it is characteristic of much of the land along the rim of the Mediterranean Sea, particularly Portugal, Spain, Italy, Greece, Israel, and Lebanon. It is also found in parts of California and Arizona, the southern tip of Australia (especially the southwest corner), the southern tip of Africa, and central Chile. Naturally, the floristic composition of such widely scattered locations is dissimilar except in general appearance. What is similar is climate.

The climate supporting the scrub forest is mild and dry. The rainfall is barely enough to support trees, usually under thirty-five inches per year. To make matters worse, the rainy season comes in winter, the most unfavorable time for growth. The summer is so dry that the vegetation is quite parched and susceptible to fire, as illustrated by the frequent brush fires in California.

Trees of the scrub forest are typically drought-resistant evergreens of the gnarled live-oak type, small and widely spaced because of limited water. Their root systems are extensive water-gathering organs, and their stems have an excellent insulation of thick cork. New crops of leaves come at the beginning of the winter-to-spring rainy season; the trees become green, and herbaceous plants spring to life and reproduce. When the dry summer comes, the plants turn drab and some become dormant. While some plants survive dry seasons by being deciduous, these evergreens do so by means of xerophytic modifications. Like many desert plants, some of them have small, thick, waxy leaves.

Underneath the trees the space may be fairly open or occupied by scrubby undergrowth mixed with grass and herbs. Most shrubs have xerophytic characteristics similar to those of the trees. Often the trees have been removed, leaving the shrubs predominating. Elsewhere the "forest" is naturally treeless, consisting of bushy thickets. Such thickets are called **chaparral** in the United States (Fig. 3–8) and **maquis** in the Mediterranean region.

Despite the arid summer climate that limits soil productivity, man has persisted in his efforts to live on scrub forest land, notably around the Mediterranean Sea. There he has found the climate favorable for growing olives and grapes. The cork oak, which is characteristic of the western Mediterranean countries, has been stripped of cork for many uses since at least 400 B.C. People who live from such land are impoverished because the yield is meager, yet they enjoy a comfortable climate. More and more people are taking advantage of the mild climate for recreation, retirement, and even permanent residency, especially near large centers of popula-

FIGURE 3–8 *A thicket of oak chaparral near Globe, Arizona. Around Los Angeles, chaparral like this often catches fire in dry weather.*

tion as in California. Where irrigation is possible, the land is often highly productive.

Animals in the Mediterranean-type scrub forest are varied but are of necessity limited in number by the available food supply. Some of the larger animals are deer, elk, bear, and cougar; smaller ones include rodents and birds.

The Mediterranean basin is a prime example of what happens under the pressure of man. Once the cradle of civilization, the area is now wasted and worn. Through the ages, deforestation for lumber and charcoal, overgrazing, and fires have taken their toll, leaving eroded soil and scrawny vegetation. In its worst form next to desolation, the land is vegetated with sparse, shrubby growth known as garrigue. Its composition is typically cistus (a small shrub), stunted kermes oak, dwarf palm, thyme, rosemary, and lavender. Maquis is a denser scrub growth found on more favorable sites. Here and there, where conditions are even better, stands of holly oak, cork oak, strawberry tree, cedar of Lebanon, wild olive, and pines may grow. Other woody trees and shrubs include laurel (used in victory crowns by the ancient Greeks), myrtle, poplar, salt cedar (*Tamarix*), acacia, pistachio, and oleander. Animal life is considerably depleted in the Mediterranean basin, but a few tigers, striped hyenas, jackals, wolves, deer, wild goats, wild sheep, and Barbary apes remain.

DECIDUOUS FOREST

Deciduous forest grows in the middle latitudes, where summers are mild and winters cold, at least severe enough to force dormancy and leaf fall (Fig. 3–9). The bulk of the deciduous forest is in the Northern Hemisphere, specifically in the United States, Central Europe, and Eastern Asia. It so happens that the area occupied by the forest roughly coincides with the areas of high human population density in which man has made the most progress.

Apparently it is more than coincidental that the two areas overlap. The polar and arid regions of the earth are obviously unproductive because of their severe climate. The forested tropical areas, though favored by temperature and humidity, do not have a corresponding potential for food production. Tropical plants produce rather continuously, but the yields of foods suitable for man are relatively low, and food storage is difficult in such areas. The yields of plants in the deciduous forest zones are cyclic, causing peak periods of food storage. Some people believe this environment—having a period of high productivity followed by a period of low or no productivity—has proved most stimulating to the cultural and

FIGURE 3–9 *A deciduous forest, primarily oak, in winter.*

economic development of man. In order to survive he was forced to plan ahead, producing and storing a surplus of food during the growing season to sustain him during the cold winter months. Possibly the necessity of looking beyond the satisfaction of immediate needs was a first step in man's progress.

Unfortunately, man's progress is offset by his destruction of the forest. In the United States much of the original forested area has been cleared. Only a few virgin stands have survived to show what the original forest was like.

The deciduous nature of vegetation is an adaptation to extreme climatic conditions. By shedding their leaves, their most metabolically active vegetative organs, plants are able to reduce their requirements and survive during unfavorable periods. Insufficient water and cold weather are the primary causes of leaf fall. A deficiency of water could result from either low precipitation or freezing. Obviously, frozen water cannot be absorbed no matter how much is present.

Leaf fall just before winter rids plants of organs that because of their shape and exposure are most vulnerable to cold damage. The remaining parts (including evergreen leaves) usually survive because they can be induced to undergo a process called winter-hardening. Exactly what changes occur in hardening or precisely what happens during a winterkill is not clearly known. It is known that hardening lowers the freezing point of protoplasm, and there seem to be two causes of winterkill—formation of ice crystals and desiccation of the tissues.

Some people believe desiccation may be more significant than freezing in winterkilling plants. According to their theory, non-protoplasmic water in the intercellular spaces freezes. In effect, this means the only diffusible water is in the protoplasm. The net movement of water would then be outward into the spaces, where it immediately freezes. Eventually desiccation kills the cell. Contrary to general opinion, the scales, hairs, and waterproofing secretions that cover the delicate tissues of winter buds have little or nothing to do with protecting them from freezing but do protect them from excessive drying.

By the time leaves fall, their photosynthetic efficiency is waning because of weather, animal, and disease damage as well as accumulated metabolites. Leaf-fall followed by spring replacement is therefore a superb way to dispose of damaged leaves (and whatever waste materials they may contain) and begin afresh with an efficient new set of leaves.

The climate of the deciduous forest biome is highly variable. For example, the temperature range is easily as much as 70° to 80°F between the low temperatures of winter and the high temperatures of summer. Rainfall varies from about 40 to 60 inches, although the extremes may reach 20 to 100 inches. Spring usually starts with an abundance of water and a rapid burst of growth. By midsummer the forest is apt to be much drier owing to faster evaporation caused by higher temperatures as well as diminishing rains. The results of these unfavorable changes can be observed in the annual growth rings of the wood—the first cells are large and the later ones small. During extremely dry periods weaker trees often die, as do those living in shallow soil where the limited water supply is quickly consumed.

Most soils in eastern America fall into two groups, the podzols that predominate in the northeastern states and the red and yellow lateritic soils found in the southeastern states. The former consists of a gray–brown topsoil overlying a brown subsoil; the latter has a red or yellow topsoil over a dark red subsoil. Humus accumulates more readily in the north, where colder weather retards the activity of organisms that bring about organic decomposition.

The large deciduous forest of eastern America has enough latitudinal spread, altitudinal range, and edaphic differences to allow for numerous community complexes. A wealth of information about the forest is available—for example, in *Deciduous Forests of Eastern North America*, compiled by E. Lucy Braun. At the risk of oversimplification, the region may be divided into three east–west belts. The northern one is rich with sugar maple associated with other species, especially beech; the middle region is dominated by oaks, commonly associated with hickory; and the southern region (on the southeastern and gulf coastal plain) is characterized by pines. Although the latter are not deciduous, they remain subclimax to oaks as a result of repeated fires or edaphic conditions. Pines also grow throughout the remainder of the deciduous forest, intermixed with hardwoods or in stands subclimax to them. They are nearly always the first trees to invade abandoned fields and often occur in almost pure stands on many drier mountainous sites, such as knobs, ridges, and peaks, or on rocky and sandy soil.

The original beauty of the forest is best preserved in public parks and forests. In the best stands the trees are large, sometimes exceeding 100 feet in height but averaging about 75 feet. The canopy is dense and continuous enough to intercept much of the light; consequently, undergrowth is sparse. There are trees in all stages of development, commonly some shrubs and vines, and a variety of herbaceous wildflowers like Solomon's seal, wild geranium, violet, May-apple, wild ginger, bloodroot, hepatica, trillium, and Jack-in-the-pulpit. From its awakening in the spring to the splashes of color it shows in the autumn, the forest is constantly changing its floral displays.

If the climax community is oak–hickory, there will probably be two or three species of each. Associated with them will be less important trees like sourwood, black gum, dogwood, and maple. In coves where moisture conditions are better, there may be basswood, tulip poplar, beech, hemlock, maple, and birch. Bottomland is likely to support willow, ash, box elder, sycamore, river birch, sweet gum, and certain species of oak, like water or willow oak.

Within the climax forest there are distinctive communities like grass or heath balds, moors, meadows, and swamps. The cypress swamps (Fig. 1–5c) of the southern coastal states and the Mississippi valley are particularly large. A few of these swamps occur as far north as southern Illinois. The cypress trees have buttressed bases for additional support and "knees," which are root extensions projecting above the water to absorb oxygen for the submerged parts. Often draped with Spanish moss, the cypresses have a somber but majestic appearance. The swamp forest is thickly inhabited by insects, fish, amphibians, snakes, turtles, birds, and mammals. In the southernmost swamps there may be alligators.

The second-growth forest, which is primarily what remains today, is ragged and ugly. The trees are predominantly culls damaged by lumbering, storms, disease, and fire. The forest is cluttered with stumps and tops as well as undergrowth that thrives when the forest floor is open to light. Allowed enough time, the forest recovers, but many of the trees that grow from stumps and roots are not sound.

In terms of variety, animal life is abundant, although at times the forest may seem unexpectedly quiet, depending on the sizes of the populations and the availability of food at a particular time. Some animals of yesteryear, like the bison, are completely gone from the deciduous forest; others, like the bear, deer, wild turkey, cougar, and bobcat, are localized. Rodents, including squirrel, chipmunk, rat, muskrat, beaver, flying squirrel, and woodchuck, are plentiful; other mammals are the bat, mole, mink, weasel, otter, fox, opossum, skunk, and raccoon. The birds occupy many niches, from that of the eagle to that of the hummingbird. Some of the reptiles found in the forest are box terrapin, fence lizard, chameleon, and snakes. Salamanders and frogs are abundant around water. Of course, every habitat has its distinctive fauna. The space beneath a fallen log, for example, would be the home of beetles, roaches, snails or slugs, spiders, sow bugs, millipedes, and centipedes. It could also include ants, termites, scorpions, salamanders, lizards, and snakes.

NORTHERN CONIFEROUS FOREST (BOREAL FOREST, NORTH WOODS, OR TAIGA)

The great **northern coniferous forest** lies between the tundra to the north and the deciduous forest to the south wherever the three biomes occur together. It is confined to large American and Eurasian landmasses in the high latitudes of the Northern Hemisphere. Because there are no large landmasses in the same latitudes of the Southern Hemisphere, such forests are not found below the Equator. The American forest extends in a wide belt from Alaska to the Atlantic Coast, dipping below the Canadian border in northern Minnesota, northern Michigan, and northern New England. Another wide coniferous belt crosses the northern part of Eurasia, beginning in northern Scandinavia and extending all the way to the Pacific Ocean, most of it in Russia. Though the species differ in America and Eurasia, the forests in the two areas are surprisingly similar in appearance.

The northern coniferous forest extends southward along north–south mountain ranges where altitude compensates for latitude in making temperatures cooler. In the Appalachian Mountains the tree species are different, yet the alpine fir–spruce stands look very much like those in the northern forest. (Fig. 3–10). Although the vegetation and climate are similar, the environment does differ in degree—the north woods are not as windy, have longer and colder winters, receive a little less precipitation, and have a different photoperiod.

The climate is best described as cold and wet—wet not so much because of precipitation, which usually varies from 15 to 40 inches per year, but because of poor drainage of the soil and slow evaporation. Much of the precipitation falls as snow and sleet.

The coniferous trees that predominate are better adapted than deciduous trees to survive the rigors of the north. Their spirelike form and short branches are well suited to the climate in that they provide a poor platform for snow and ice, which would break the limbs if it accumulated. The growing season is short, followed by a severe winter during which water is frozen and unavailable for absorption. The needlelike leaves of the evergreens are thicker (small surface area), with sunken stomates and thick cuticles that retard transpiration, a process that can dehydrate and kill the trees during critical periods when winds are heavy or the ground is frozen. Their short needles are definitely superior for shedding snow to the long needles on some trees. Even the jack pine, which has the longest needles of any of the widespread species found in the coniferous forest, has shorter needles than many pines growing farther south, where snow seldom falls.

The terrain of the northern woods was glaciated during the recent Pleistocene ice age, leaving the surface scoured in some places and littered with deposits in others. The established drainage pattern was completely disrupted, leaving the countryside dotted with lakes and swamps. The patchwork pattern of bare rock and watery habitats

FIGURE 3–10 *An alpine fir–spruce forest along the Appalachian Trail on the Tennessee–North Carolina border. The fir* (Abies fraseri) *and spruce* (Picea rubens) *are reminiscent of the northern coniferous forest.*

left by glaciation interrupts the continuity of the forest and makes the region difficult to develop or exploit. A look at a map of Canada, where much of the forest is located, draws attention to the dearth of settlements and roads. The obstacles to building roads and the rigors of the climate at such high latitudes keep the country untamed. Under such climatic stress, farming is definitely unattractive. Logging for lumber and pulp, mining, hunting, and fishing remain the most promising possibilities for economical use of the area.

The mature soil type of the northern coniferous forest is a podzol whose upper layer (Horizon A) is ash-colored and resembles even-grained sand. Overlying it is a thick accumulation of organic matter that decays slowly because of the cold climate. Water percolating through the humus becomes acid and washes out basic substances and fine-grained particles as it passes through the topsoil. Underneath the topsoil is a compact, dark brown subsoil (Horizon B), where leached material is deposited. Together the two layers are shallow, usually less than two feet in thickness.

The predominant trees of the American coniferous forest are the spruces, white and black, which are transcontinental in distribution. White spruce grows best on well-drained soil, whereas black spruce flourishes in the soggy soil prevalent throughout the forest. A close associate of the two species of spruce in the eastern two-thirds of their range is balsam fir, a similar life form. These three trees point skyward like dark green, jagged spires and give the coniferous forest its special aspect. Viewed from a distance, their narrow, pointed tips give the impression that they are more widely spaced than they really are. The shrubs grow up to 4 feet high, low herbaceous plants up to a foot or so, and mosses and lichens at the ground level. Sphagnum forms a thick carpet on the forest floor, especially in a low-lying forest.

A bog in a coniferous forest is known as a **muskeg**. It may be completely open or contain a few scattered or stunted trees. It may be filled with ericaceous (heath) shrubs 3 or 4 feet high (Fig. 3–11) on a carpet of sphagnum, or there may be open beds of sphagnum. In other muskegs there may be a central pool of water encircled by zones of vegetation that grow progressively toward the center as the depression fills. These zones, from the edge of the water outward, usually consist of a border of sphagnum and other aquatic plants (including sedges and grasses), an intermediate belt of ericaceous shrubs, and the lowland forest trees (black spruce, tamarack, and eastern white cedar) abutting the upland forest.

The northern coniferous forest is good game country and offers the promise of developing into a frontier recreational area. Bear, moose, caribou, deer, ducks, and geese are plentiful for the hunter, as are trout, bass, pike, pickerel, and muskelunge for the fisherman. Other animals include beaver, muskrat, fox, weasel, mink, marten, wolf, wolverine, and rabbit.

FIGURE 3–11 *A heath bog in a northern coniferous forest in Wisconsin. Beneath the shrubs is an almost solid mat of sphagnum. The trees surrounding the bog are black spruce* (Picea mariana).

FIGURE 3–12 *The relationship of altitudinal and latitudinal vegetation zones. (From C. Leland Rodgers,* Essentials of Biology, *©1967, Barron's Educational Series, Inc., Woodbury, New York.)*

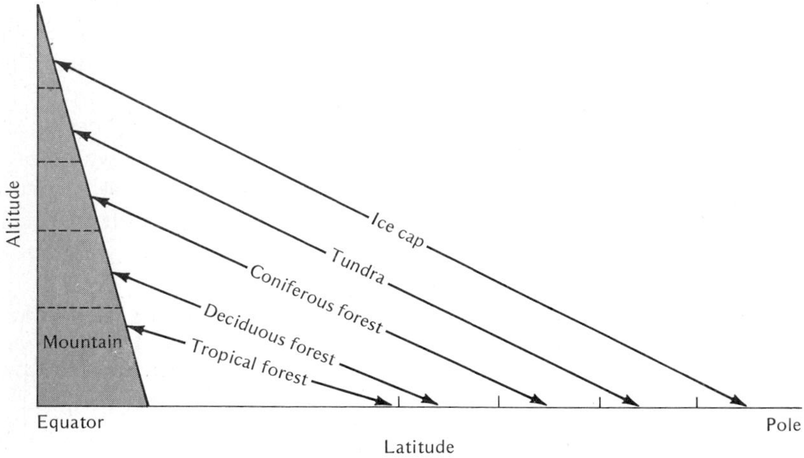

FIGURE 3–13 *Timberline of the Colorado Rockies. (a) Timberline, the boundary between the coniferous forest and alpine tundra. Often this is a sharply defined boundary. (b) Remains of dead trees above timberline. The position of the timberline fluctuates in response to long-term climatic changes.*

a

b

MONTANE ZONATION

On steep mountain slopes there is a substantial environmental gradient that causes the development of distinct life zones corresponding to biomes. From bottom to top, the zones more or less correspond to the sequence of biomes from the equator to the poles (Fig. 3–12). The boundary between zones may or may not be sharply delimited (Fig. 3–13). The zone at the foot of the mountain would normally be the climax community for that particular geographic region, and the zones above it would be like biomes of colder regions. For example, a high mountain at the equator could have a tropical forest at the bottom followed by zones of deciduous forest, coniferous forest, tundra, and ice.

Of primary influence on montane zonation is the temperature gradient, which changes approximately 1°F for every 333 feet in elevation. Coupled with the temperature gradient is a humidity gradient. Since colder air at higher elevations cannot hold as much moisture as warmer air at lower elevations, precipitation is higher. Sometimes local conditions exist in which updrafts of moisture-laden warm air along a mountain slope produce large amounts of rainfall as they cool. Although the temperature and moisture gradients are the primary causes of montane zonation, their effects are modified by several other factors such as exposure, slope, or soil conditions.

REFERENCES AND FURTHER READINGS

Billings, W. D. 1968. *Plants, Man and the Ecosystem*. Wadsworth, Belmont, Calif.

Eyre, S. R. 1968. *Vegetation and Soils, a World Picture*. 2d ed. Aldine, Chicago.

Farb, P. 1961. *The Forest*. Time-Life, New York.

Gourou, P. 1966. *The Tropical World*. 4th ed. Wiley, New York.

Hanson, H. C., and E. D. Churchill. 1961. *The Plant Community*. Reinhold, New York.

Krebs, C. J. 1972. *Ecology*. Harper & Row, New York.

Kucera, C. L. 1973. *The Challenge of Ecology*. Mosby, St. Louis, Mo.

Neal, E. 1964. *Woodland Ecology*. Harvard University Press, Cambridge, Mass.

Oosting, H. J. 1956. *The Study of Plant Communities*. 2d ed. Freeman, San Francisco.

Preston, R. J., Jr. 1961. *North American Trees*. 2d ed. M.I.T. Press, Cambridge.

Richards, P. W. 1952. *The Tropical Rain Forest*. Cambridge University Press, Cambridge.

Chapter 4
Grasslands

CAUSES

Wherever grasses dominate the countryside, they betray the presence of drought as a major ecological influence; many grasses are well equipped to live where trees cannot. While forests generally require more than thirty inches of rainfall per year, grasslands can thrive on less. Although the annual rainfall in grasslands may vary from ten to eighty inches (more commonly ten to thirty inches), it is by no means equally distributed throughout the year. The drought period ranges from one to three months and occasionally lasts six months or more. During that time the treeless, windswept expanse looks dry and brown.

Besides the quantity and seasonal distribution of precipitation, other causes of grasslands are soil type, grazing, fire, and land use by man. Probably much of the grassland is climax, though parts of it might develop into something else if certain factors like grazing or fire were eliminated.

CHARACTERISTIC LIFE FORM

The grass life form is a highly successful adaptation to the environment just described. Most grasses are perennials that can lose the majority of their above-ground part through death or removal and yet survive on what remains near or below the surface. Grass leaves, growing from the bottom, can be cropped anywhere above the growth zone and continue to grow from their bases. Even when grazed close to the ground, most grasses send up new shoots. Fires do little damage, since they occur in the dry season and burn mostly dead leaves and stems. Fortunately, grasses can reproduce without the aid of pollinating insects, which are relatively scarce in grassland biomes. They depend on wind pollination or propagate vegetatively.

Grasses likewise have ways to make the most of the arid conditions under which they live. Their root systems are profusely branched and grow close to the surface, where they can quickly absorb even small quantities of rain. In many species most of the root system is found in the upper two feet of soil. Although they are efficient in obtaining water, most grasses are also distinguished by the capacity to survive long periods of drought and dehydration of their tissues without serious injury.

Many have assumed that various modifications of the structure of grass leaves allow the plants to be more efficient in water utilization. Experiments have shown, however, that when sufficient moisture is available such plants frequently have transpiration rates equal to or even surpassing those of plants found in moister habitats. Their chief advantage, then, is the ability to endure desiccation. Under drought conditions some grasses can roll their leaves and reduce transpiration to 5 to 10 percent of the normal rate. The stomates, occurring almost entirely on the upper surface of the leaf, are enclosed as the leaf rolls up. They are also covered by a thick, waxy cuticle that reduces transpiration from exposed epidermal tissues. In the desiccated condition the plant looks nearly dead, but it can revive quickly when water becomes available. This phenomenon of rolling and unrolling leaves explains why a parched lawn or field may revive quickly after a shower.

ANIMAL ADAPTATIONS

Without trees for shelter and protection, grassland animals have to make the most of what they have—wide open spaces. They are in general well suited for either subterranean or surface existence. If they are primarily surface creatures, they are often long-legged and swift. This is particularly true of the large herbivores, whose primary defense is flight from predators and fire. Their horny hoofs are efficient for straightaway travel on hard, sunbaked ground. They are also good for traveling long distances to find food or water as the animals often have to do during the dry season.

The large primary consumers are the grazing herbivores, which have grinding teeth for mastication and digestive tracts suitable for processing bulky plant foods. Their teeth contain alternating bands of hard enamel and softer dentine. The abrasive action of teeth against teeth, as well as sand grains inadvertently taken in with grass, wears away the softer dentine at a faster rate and leaves sharp cutting ridges of enamel. The alimentary tract of herbivores is relatively larger than that of carnivores. Its size is essential for processing the large quantities of less nutritious vegetable matter used for food. Even the large body size of herbivores may be required to handle bulky food—seedeaters and carnivores are naturally smaller.

To the surprise of most people, a study of a grassland area in Colorado showed that the ground was teeming with mites (more than on forested land), which are probably substantial primary consumers. An expert working in the field reported that these mites may be a greater grazing force than all of the other herbivores combined.

Besides the obvious adaptations to get food and flee from danger, grassland animals have a variety of other specializations. Many of them have an inclination for gregariousness, by which they achieve strength in numbers. With hiding places few and far between, larger animals tend to roam in herds and smaller ones to excavate labyrinths of underground burrows. Even the nongregarious burrowing owl goes underground for nesting. Other birds, like the horned lark, construct nests on the ground. A considerable number of the residents of grassland have protective coloration.

CONVERGENT EVOLUTION

Wherever the same types of habitat exist throughout the world, the occupants often have similar habits and look alike regardless of their ancestry. They are called ecological equivalents. What they have done is to adapt to prevailing conditions by the process of evolution. In the language of biologists they fill the niches, meaning they occupy unique relationships with respect to all other organisms with which they are associated. In toto, their requirements for food, shelter, and space are never exactly like those of any other organism. Neither is their predator-prey relationship the same. The widely scattered grasslands are good places to look for examples of ecological equivalents.

Because of their food demands, the herbivores are particularly identified with grasslands. They may be divided into running or jumping types. Living with the large herbivores are carnivores and burrowing rodents. Table 4–1 shows examples of ecological equivalents for each continent where large grasslands occur. Only one example from each continent is used in order to keep the comparisons simple.

UTILIZATION

Grasslands are good areas for agriculture and grazing. When America was first explored, the "sea of grass" seemed like wasteland

TABLE 4–1 *Ecological-equivalent mammals native to grasslands*

	North America	South America	Africa	Eurasia	Australia
running herbivores	pronghorn antelope	Pampas deer	zebra	wild horse	—
jumping herbivores	jack-rabbit	mara	spring-haas	jerboa	kangaroo
carnivores	coyote	South American jackal	hyena	Tibetan wolf	Tasmanian "wolf"
burrowing rodents	prairie dog	Patagonian cavy	African ground squirrel	hamster	wombat

compared to the forested East. Now, nearly all but the drier western part has been put to the plow and is covered with fields of grain. Where practical, productivity of large tracts has been increased by irrigation from wells and streams. Even some of the drier western land, which is marginal for agriculture, is strip farmed (Fig. 4–1). The

FIGURE 4–1 *Strip farming in dry area of northern Montana. Half of the strips are planted in wheat while the other half remain fallow. In the following year the fallow strips are planted and the cropped strips left fallow.*

FIGURE 4–2 *Rangeland poorly suited for agriculture because of low rainfall. Here, in northeastern New Mexico, the dominant vegetation consists of grasses like buffalo grass* (Buchloe dactyloides) *and blue grama* (Bouteloua gracilis) *and succulents like cholla and other cacti.*

remainder is rangeland (Fig. 4–2). The entire expanse has been so disturbed that fragments of the original biome are almost impossible to find.

SAVANNA (INCLUDING SOUTH AFRICAN VELD)

The **savanna**, a tropical biome, is an extensive grassland often interrupted by a few scattered trees, or it may be a parklike mixture of grasses and trees. It is grazed and browsed by herds of roaming herbivores, which in turn provide food for tag-along carnivores. Savannas are located above and below the tropical rainforest of South America and Africa as well as in other, smaller tropical and subtropical areas of Central America, Madagascar, India, southeastern Asia, northern Australia, the East Indies, and the West Indies. These locations are warm all year around and subject to frost only on very rare occasions near their temperate-latitude boundaries.

The savannas are covered with coarse, tall grasses from six to twelve feet tall. Most plants grow singly rather than clumped. During the wet season the tall grasses, spread across extensive plains, are an impressive sight. When dry weather comes the grasses turn brown as the tops begin to die and slump to the ground. Besides the dominant grasses, there are a variety of legumes and other herbaceous plants. Trees, if present, are typically small, thorny, and small-leaved. Especially in parts of Africa, they are often flat-topped. In both Africa and South America there are unusual trees with barrel-shaped trunks.

Most people visualize the savanna in terms of the famous Big Game Country of Africa. The African grassland is the most extensive in the world, occupying perhaps one-third of the continent. In South Africa it is called the **veld**.

Herbivores of the Big Game Country include elephants, giraffes, zebras, rhinoceroses, wart hogs, bush pigs, buffaloes, and a host of antelopes such as impalas, hartebeests, wildebeests, elands, and gazelles. Many of them are preyed upon by cheetahs, leopards, and lions, all members of the cat family. Lurking at a distance are the scavenging vultures, jackals, and hyenas, awaiting their turn to clean the carcass of the kill. Other animal life is found in the tropical scrub forest (Chapter 3), a biome adjoining the savanna. Of the invertebrates, the mound-building termites are significant.

Open grassland and the parklike mixture of grass and long-leaf pine covering the southern coastal plain of the United States are inappropriately called savanna, if for no other reason than that they are not tropical. The same is true of temperate grasslands in other places. In the United States only the Everglades region of Florida is similar to the savannas. At least it has a tropical aspect and seldom suffers frost.

The Everglades region is, however, unique in several respects. Actually, the tract of land known as the Everglades is a broad, shallow drainage trough that connects Lake Okeechobee to the sea—in reality a river bed clogged with tall saw grass (*Cladium jamaicense*). Contrary to its name and superficial appearance, the distinctive plant is not a true grass but a sedge. Because of its appearance and topography, the Everglades has been called a "river of grass," a "river" as much as ninety miles wide. Water, normally six to nine inches deep, covers the glades for less than five months of the year. Before man interfered by diverting water to agricultural and other usage, water covered the glades for nine months each year. The glades are very low, approximately ten feet above sea level. Most of the fifty to sixty inches of rainfall comes from May to November, and the area is flooded with water slowly draining to the sea. Some organisms avoid the flooding by building nests in the vegetation

FIGURE 4–3 *An ant nest attached to saw grass* (Cladium jamaicense) *in the Everglades. The nest is high enough above the ground to protect it from flooding.*

above the water level (Fig. 4–3). During the dry season water is confined to the deep, seemingly aimless channels and scattered small pools. As on the true savanna, the sawgrass becomes parched and subject to fire. Scattered through the glades are islands of trees called hammocks (Fig. 4–4).

FIGURE 4–4 *The Everglades of Florida. The low ground, covered with saw grass, surrounds hammocks, or islands of trees located on slightly higher ground.*

PRAIRIES (INCLUDING ARGENTINE PAMPAS) AND STEPPES

Grasslands of the temperate latitudes were once great expanses of grasses interrupted only infrequently by small dissimilar communities on wetter or drier sites. Because of their lushness, they harbored herds of grazing animals. The original vegetation has been all but destroyed in man's exploitation of the region by the grazing of domesticated animals and by cultivation.

Temperate grasslands are divided into prairies and steppes, depending on the size of the grass, which in turn is related to moisture conditions. True prairie grasses are thick and are usually about two to three feet tall (Fig. 4–5), but they may reach a height of six to eight feet; short steppe grasses are wider spaced and about five to ten inches tall (Fig. 4–5). The larger prairie grasses grow adjacent to forest lands where soils are moister; steppe grasses grow adjacent to deserts where it is drier. From the eastern to the western extreme of the American grassland, grass height closely parallels a moisture gradient.

The North American grassland stretches across the heart of the United States for 800 to 1000 miles and north–south from Canada to Mexico for over 2000 miles. The original prairie occupied the midcontinental lowland and roughly coincides with the present-day corn belt. Steppe originally covered the Great Plains, the long slope from the foot of the Rockies to the floodplain of the Mississippi. The eastern part of this area is now planted mostly in wheat and grain sorghums, and the drier western part is principally ranchland. The transition from prairie to steppe is gradual but occurs fairly close to the 98th meridian, which runs through central Kansas and Nebraska. The transition zone is often called the **mixed prairie** (Fig. 4–5). East of that strip annual rainfall is twenty or more inches whereas west of it the rainfall is generally less than twenty inches.

Some temperate grasslands are located on every continent, though most of this biome is found in central North America and Eurasia. Smaller Southern Hemisphere locations are the pampas of Argentina and Uruguay, New Zealand (tussock grassland), Australia, and Africa.

Grass has probably encroached on forest land because of animal pressures and fire, which are normally much more destructive to trees. Large herds of herbivores, entering the forest edge from the grassland, probably damaged the older trees by browsing and trampling, and destroyed the younger trees that would have replaced them. Once grass became established, its thick sod was an unfavorable site for tree seeds to sprout or seedlings to survive. Other possible reasons why grasses grow where trees would be expected are prolonged droughts and high temperatures that accelerate evaporation beyond what trees could normally tolerate.

Animal pressures and fire probably extend the boundaries of grassland into forest land; the reverse should be true when they are removed. In particular, the control of fire has permitted certain trees to invade grassland to the point where they are serious competitors of the grasses. A case in point is the mesquite tree, which is taking over some areas like a weed. Periodic fires would have killed the seedlings and young mesquite trees without damaging the grass. In an effort to reclaim the land for grass, many ranchers are trying to kill the trees by such techniques as spraying them with herbicides from the air or uprooting them by bulldozing or dragging a heavy chain between two bulldozers (the chain catches the tree and pulls it out of the ground). The cost of clearing mesquite from grassland is often higher than the value of the grass.

In general, the composition and structure of grassland soils are influenced most by aridity. This means their profiles are relatively shallow compared to that of soils developed where the rainfall is heavy. As the water moves down through the soil, it takes the fine particulate matter and dissolved calcium from the top zone and deposits them at a lower level. The calcium accumulates as a hardpan except where the rainfall is highest. In chernozem soils, a well-known grassland type, calcium is deposited at a depth of about three feet. Brown (chestnut) soils developing under more arid conditions have a somewhat more shallow profile, and prairie soils developing where rainfall is higher have a deeper profile. Prairie soils do not have the calcareous hardpan typical of grassland soils in more arid regions.

As the name implies, the predominant grassland life form is the drought-resistant grasses. Usually five or six species provide 95 percent of the food for grazing. Some types of American grasses are buffalo grass, grama, bluestem, wiregrass, and wheatgrass; some types found in the extensive grasslands of Russia are fescue, koeleria, and feather (needle) grass. Common **forbs** of the American grassland are legumes and composites. Trees, when present, ordinarily grow along streams (Fig. 4–6) or are planted in shelter belts (Fig. 4–7) around homesites. The most obvious type of tree along streams is the cottonwood; those in shelter belts are often elm, box elder, ash, Russian olive, and poplar.

FIGURE 4–5 *The grasslands of the Great Plains. (a) True prairie. In addition to a number of species of forbs, the true prairie is characterized by midgrasses, including bluestem* (Andropogon scoparius), *side-oats grama* (Bouteloua curtipendula), *western wheatgrass* (Agropyron smithii), *porcupine grass* (Stipa spartea), *switch grass* (Panicum virgatum), *and tall dropseed* (Sporobolus asper). *Much of the true prairie has been eliminated by cultivation. (b) Short-grass prairie. The major species of vegetation found on the short-grass prairie (steppe) are buffalo grass* (Buchloe dactyloides), *blue gama* (Bouteloua gracilis), *and red threeawn* (Aristida longiseta). *(c) Mixed prairie. The mixed prairie, between the true prairie and the short-grass prairie, features both sod-forming and bunch grasses. Major species include needlegrass* (Stipa comata), *western wheatgrass* (Agropyron smithii), *bluestem* (Andropogon scoparius), *side-oats grama* (Bouteloua curtipendula), *blue grama* (Bouteloua gracilis), *and purple threeawn* (Aristida purpurea). *(Photos courtesy of G. W. Tomanek.)*

a

b

c

Animal life on the original grasslands of America was drastically altered by man as he claimed the land. Until recently the bison was king of the North American plains, but this animal is now relegated to a few small, protected spots. Between 1820 and 1894, when a bill was passed to protect bison, the population dwindled from many millions almost to extinction. Today it has grown to about 10,000 animals. Surviving and thriving, at least under present conditions, are the deer and pronghorn antelope (not a true antelope). The deer is primarily a browsing animal; the pronghorn grazes and

FIGURE 4–6 *Cottonwoods* (Populus) *along the banks of a dried-up stream in Canyon de Chelly, Arizona. Cottonwood trees are characteristic of stream banks in arid regions throughout the West.*

FIGURE 4–7 *A shelter belt (end view) of trees in the Great Plains. Shelter belts are commonly planted in the Great Plains to break the force of prevailing winds.*

browses. The pronghorn in particular is admirably suited for grassland existence. It is fleet (can run a mile a minute on the getaway) and has keen wide-angle vision, both qualities needed to escape predators. Typical of grassland animals, it is gregarious. A white patch of hair on its rump gives it the ability to signal danger to others of the herd. Among the large herbivores on Eurasian grasslands are the wild ass, wild horse, and antelope (saiga).

The chief surviving mammalian predator on the American scene is the coyote. Like the bison, the wolf used to be a prominent member of the community but has succumbed to the attacks of man. It is still present in significant numbers in parts of the Eurasian grasslands.

Smaller mammals are numerous, each playing its unique role in the biological economy. The jackrabbit, now in smaller numbers than in former times, has quick-sprint ability and large ears to help detect danger. Like the jackrabbit, the prairie dog (Fig. 4–8) is considered a nuisance and has been relentlessly attacked. It lives in

FIGURE 4–8 *A prairie dog* (Cynomys) *near its burrow in North Dakota. Prairie dogs live in "towns" consisting of a maze of underground burrows.*

villages consisting of a maze of tunnels and outlets. Other small grassland mammals are gophers, badgers, porcupines, skunks, ground squirrels, cottontails, foxes, rats, mice, squirrels, and muskrats. Some Eurasian ones are ground squirrels, marmots, voles (mouse- or ratlike rodents), and jerboas (nocturnal, jumping rodents).

Among the birds, the ground and soaring types fare best in grasslands. Typical ground birds are prairie chicken, pheasant (introduced from Asia), and quail; among the soaring birds are vultures, eagles, and hawks. Around ponds and lakes, ducks and other waterfowl abound. Other abundant bird species include mourning doves, redwing blackbirds, horned larks, meadow larks, crows, and brown thrashers. Year-round residents are few because of the scarcity of food and limited shelter. Migratory birds following the Mississippi and central flyways pause as they cross the midcontinental grasslands.

REFERENCES AND FURTHER READINGS

Billings, W. D. 1968. *Plants, Man, and the Ecosystem.* Wadsworth, Belmont, Calif.

Hanson, H. C., and E. D. Churchill. 1961. *The Plant Community.* Reinhold, New York.

Humphrey, R. R. 1962. *Range Ecology.* Ronald, New York.

Odum, E. P. 1971. *Fundamentals of Ecology.* 3d ed. Saunders, Philadelphia.

Weaver, J. E., and F. W. Albertson. 1956. *Grasslands of the Great Plains.* Johnsen Publishing, Lincoln, Neb.

Chapter 5
Deserts and Tundras: Biomes of Extreme Environmental Stress

BIOMES IN EXTREME ENVIRONMENTS

Water and temperature are the primary determinants of biomes. Either scarcity of water or extremes of temperature can limit metabolic activities, which in turn limit growth. For the most part, the results of such limiting factors on plant life are small size and wide spacing. With plant life limited, the size of animal populations is restricted by lack of food.

Deserts and tundras are two biomes subject to extreme climatic conditions. Though quite distinct, they have some characteristics in common. Indeed, some people have called the tundras frozen deserts, alluding to their desolation and lack of absorbable water (frozen for much of the year).

Needless to say, both deserts and tundras are inhospitable to man. He finds both of them unproductive and uncomfortable. Natives tend to eke out a living in nomadic existence, wandering here and there in search of food. They are weather hardened and accustomed to deprivation.

DESERTS

Deserts are often defined as areas receiving less than ten inches of rainfall per year. The trouble with such a simple criterion is that it does not take into account the seasonal distribution of rainfall or the amount retained by the soil. Effective rainfall could amount to considerably less than the records show, because it may fall faster than it can be absorbed or may fall on steep or impervious surfaces and run off. In either case the runoff is lost from the immediate area. Subtract evaporation and transpiration losses from effective rainfall and little is left to sustain life.

A better criterion for determining what is desert is the amount and kind of life that is present. Life is virtually absent or sparse in a setting where boldly exposed features of the geosphere seem to dwarf the struggling plants. Though the larger perennial plants are almost always succulent (Fig. 5–1) or woody, they are noticeably spaced, with a considerable amount of intervening open surface area. The wide spacing reduces competition for what little water is available.

The deserts of the world occur more or less in two belts astride the tropics of Cancer and Capricorn. Because of global wind patterns, the prevailing winds are drier than those farther north or south. To make matters worse, much moisture is lost before it can reach inland areas or the leeward sides (rainshadows) of mountain ranges. Consequently, deserts are usually large inland landmasses or strips along the western fringes of continents. While all deserts are dry, they may also be classified as hot or cold.

Contrary to popular opinion, most deserts are not covered by shifting sand. The sandiest one is the Arabian Desert, which is about

FIGURE 5–1 *Idria, a strange-looking succulent found in Baja California, Mexico. This plant is of the family* Fouquieriaceae *and is not a cactus.*

one-third dunes. The Sahara, often considered sandy, has only about 10 percent of its area covered with sand. Actually, most desert soil is coarse-grained, pebbly, or rocky and supports vegetation. Beneath it but close to the surface is a hardpan composed mainly of calcium carbonate. In low places such as dried-up lake beds the surface will be encrusted with mineral salts.

Desert soils are coarse because there is less chemical weathering; the finer dust and sand particles are blown elsewhere, leaving the heavier pieces behind. On hot days several dust devils (whirlwinds) may be seen simultaneously in the distance. Sometimes sandstorms are severe enough to sandblast the paint from cars and "frost" glass. During large windstorms fine dust particles may be transported beyond the border of the desert and deposited as loess. The heavier pebbles that remain after the dust and sand have been blown away may be of sufficient size and number to settle into a firm, smooth, neatly arranged mosaic, which is known as desert pavement and looks very much like gravel pavement.

Rainfall in the desert is often very low and/or concentrated in a short period between long rainless periods. Evaporation rates regularly exceed rainfall rates. Sometimes rain starts falling and evaporates before reaching the ground. Rainfall is lowest on the Atacama Desert of Chile, where it averages less than half an inch and some years are rainless. The interior of the Sahara also receives less than an inch a year. Rainfall in the American deserts is higher—almost eleven inches at Tucson, Arizona.

Temperatures of deserts exhibit wide daily extremes because the atmosphere contains little humidity to block the sun rays on their way in or to reduce the reradiation of heat from the atmosphere into space. Desert surfaces receive a little more than twice the solar radiation received by humid regions (90 percent compared to 40 percent) and lose almost twice as much heat at night. On the Sonoran Desert the summer temperature can reach a little above 100°F during the day and be cool enough at night to require a blanket for sleeping—sometimes the difference is more than 50°. Night cooling helps both plants and animals by reducing evaporative losses due to transpiration, sweating, and breathing. Moreover, condensation of dew caused by night cooling may equal or exceed the rainfall received by some deserts.

There are four major desert areas in North America—the Chihuahuan, Sonoran, Mojave, and Great Basin. The Chihuahuan and Sonoran deserts are located in both Mexico and the United States and are named after Mexican states. The Chihuahuan Desert lies east of the Sierra Madre Occidental of Mexico and extends into southern New Mexico and southwestern Texas. The Sonoran Desert is west of the Sierra Madre Occidental (including Baja California) and extends northward into southeastern California, southern Nevada, the southwestern corner of Utah, and a small portion of western Arizona. The Mojave Desert is restricted primarily to southeastern California and adjacent small areas in western Arizona and southern Nevada. The Great Basin Desert, unlike the preceding ones, is a cold desert located between the Sierra Nevada and Cascade ranges to the west and the Rockies to the east. Together the deserts occupy about 5 percent of North America, which is a small portion by contrast to that covered by the Australian Desert, which occupies about 44 percent of that continent.

North American deserts can be distinguished from one another by a few conspicuous species. The Sonoran (Fig. 1–3) features the columnar cacti, the Mojave (Figs. 5–2 and 11–5) the Joshua tree, the Great Basin (Fig. 5–3) the sagebrush (*Artemisia*), and the Chihuahuan an abundance of yucca and prickly pear, neither of which is confined to this desert alone.

FIGURE 5–2 *Joshua tree* (Yucca brevifolia) *characteristic of the Mojave Desert. This scene is in western Arizona.*

FIGURE 5–3 *Sagebrush* (Artemisia) *of the Great Plains. This scene is in northern Wyoming.*

The scarcity of water and high daytime temperatures that speed evaporation make it imperative for plants to possess protective adaptations. Some, like the saguaro cactus, have extensive root systems close to the surface where they can make the most of small showers. All of the succulents can take in large quantities of water when it is available and store it for future use. The surface of the saguaro, for example, is corrugated with longitudinal ridges and grooves. When water is available, the stem swells so that the grooves are shallow and the ridges far apart. As the water is used, the stem shrinks so that the grooves are deep and the ridges close together. Other plants, like the mesquite, are claimed to have root systems deep enough to tap the groundwater reserve as much as 100 feet below the surface.

The leaves of desert plants are replete with water-conserving characteristics. They tend to be small and thick, and covered with a thick cuticle. In the cacti the leaves are reduced to ephemeral vestiges, and photosynthetic activity is restricted to the stems. In plants like the ocotillo and palo verde, the transpiring surface is reduced by leaf-fall during dry periods. The ocotillo quickly regrows a new set of leaves when rainfall recurs. Some desert plants open their stomata only at night when evaporation rates are lowest.

Other plants die back all the way to the ground and grow back from dormant subterranean parts when conditions are more favorable. The spiny nature of many desert plants provides added protection in a particularly hazardous environment. Besides discouraging animals, large numbers of spines, like those of the barrel cactus, shade the surface enough to significantly reduce transpiration. The same may be true of the hairs on the woolly plants that are so abundant in deserts. Many plants have silvery or glossy leaves, allowing them to reflect more radiant energy. Also, desert plants often have an objectionable odor or taste.

An entirely different solution to desert survival is that followed by annual plants. Many grow rapidly and set seeds in a much shorter period than most annuals do elsewhere. Seeds are produced in abundance and may remain dormant for years before the season is right for germination. Some have hard seed coats that need scarifying, which can be accomplished by the abrasive action of wind- and waterborne sand and gravel. Others have inhibiting chemicals in the seed or fruit coats that require more than a light shower to dissolve and wash them away. Requiring a sizable rain to remove inhibitors keeps the seed from germinating before there is enough moisture available to sustain the seedlings. Beyond that, some seeds still do not germinate unless the temperature is right.

Desert animal life does not seem nearly so specialized as plant life. Most animals are very much like animals elsewhere. They simply adjust their schedules by staying inactive in protected hideaways during the heat of the day and coming out to forage at dusk and dawn or at night, when the desert is cooler. During daylight hours insects will move around twigs to keep on the shady side; jackrabbits will follow the shifting shadow of a cactus or shrub. Naturally, many animals seek protection in underground burrows where they are insulated from both heat and aridity. For example, it has been reported that the temperature of a burrow eighteen inches below the surface is 61°F when the surface temperature is 150° (Mojave Desert).

Specialized adaptations of desert animals usually have something to do with adjustments to heat, water, and locomotion. In what amounts to a kind of summer hibernation, some animals, like the spadefoot toad, estivate during hot, dry periods. The spadefoot toad seals itself in a burrow with a gelatinous secretion and remains inactive for eight or nine months until a heavy rain occurs. Amphibians and other animals that pass through aquatic larval stages have accelerated life cycles, which improve their chances of reaching maturity before an unfavorable turn in the weather can terminate the cycle. Quite a few animals have dormant eggs or pupae that do not resume development until the surroundings are suitable. This is especially true of insects. Most spectacular, however, are the fairy shrimp and a few of their close relatives. Their eggs remain dormant in the crusty, parched floors of playas for years (twenty-five years in one case recorded in the Mojave Desert) and hatch in the temporary pools of water that stand after a rain.

The evolution of water-conserving and storage adaptations is a distinctive feature of some desert animals. A few do not drink water; instead, they depend on the water of metabolism, which they conserve. A good example is the American kangaroo rat (not a true rat), so named because it hops around on enlarged hind legs and is balanced by a large tail. The major item in its diet is dried seeds, whose water content is only about 4 percent. Metabolic water produced during respiration is made available to the organism.

A single kangaroo rat metabolizes about two ounces of water in five weeks. This would not be enough if it were not conserved by the rodent itself. The rat is able to do this because it does not sweat, nor does it use much water to flush waste from its body. Because of the low protein content of its diet, nitrogenous waste is minimal; consequently, the animal seldom urinates. The kangaroo rat's urine is very concentrated—about 70 percent water compared to 92 percent in that of humans. The rat's water loss through evaporation from breathing is drastically reduced during the day because it seals itself in an underground burrow where the relative humidity is much higher than outside and the moisture from breathing is retained within the enclosure. The animal forages at night, when water loss to a cool atmosphere is considerably less than it would be in the heat of the day.

The camel, like the kangaroo rat, makes use of metabolic water. It can go at least a week without water or food. Fat is stored in its hump and used for food and water when necessary. During periods of water deficiency, urination and sweating are reduced, and breathing and panting are slow. Lack of water intake and reduced cooling processes are accompanied by loss of weight and an increase in body temperature to as much as 9°F above normal. The camel quickly returns to normal when food and water become available again.

Besides the kangaroo rat and the camel, which are well known by man, other animals are well fitted for desert existence. The gemsboks and oryxes of East Africa take no water, and zebras go from one to three days without drinking. The desert tortoise of the Sonoran and Mojave deserts not only gets water from food but can store enough to last through the hot season. It stores the reserve in two sacs underneath its carapace.

Other adaptations of desert animals take unusual forms. In addition to capturing sound waves, the large ears of jackrabbits and kit foxes probably dissipate heat much as the fins of a radiator do, by increasing the surface area exposed to the air. The sidewinder rattlesnake of North American deserts and the sand viper of the Sahara have a peculiar looping motion that gives them good traction on sand. A fringetoed lizard inhabiting sand dunes can "swim" in the loose sand, aided by the increased contact the fringes provide. Some snakes and lizards that "swim" in sand have nostril valves to protect them from breathing sand grains into their lungs.

The quantity of animal life in deserts is related to the availability of plant foods, both vegetative parts and seeds. Although total productivity of desert plants is comparatively low, seeds are produced in enormous quantities by annual plants and support a large population of rodents like the pack rat (Fig. 5–4) and kangaroo rat. Besides the rodents, there are many insects, reptiles, and birds—far more than most people suppose. Some of the larger American desert mammals are rabbit, skunk, peccary, coati, kit fox, gray fox, badger, bobcat, pronghorn antelope, deer, and coyote. Niches in deserts the world over are filled by similar animals; in Australia they are filled by specialized marsupials.

Unproductive as they are, deserts provide a living for a surprising

FIGURE 5–4 *Pack rat burrow under a cholla plant. The spines on the cholla sections around the mouth of the burrow provide a measure of protection.*

FIGURE 5–5 *An irrigation pump beside a cottonfield near Phoenix, Arizona. The pump is withdrawing fossil groundwater. The cottonwood beside the ditch was killed by girdling in order to conserve water that the tree would otherwise have used.*

number of people. Existence in the desert is of necessity simple and primitive. Inhabitants use wild plants for food and keep animals like sheep, goats, camels, or cattle. Sometimes they practice a little dry farming by planting a small garden plot or orchard at the base of a large slope or in a ravine, where runoff from rains moistens the soil. Homes are built of native materials like grass, adobe, poles, skins, or animal hair (tents). If people do not have to be nomadic to obtain their food, adobe is a good building material; adobe homes are cooler in summer and warmer in winter.

People are moving into the deserts in greater numbers and building modern cities and farms. Farming is prosperous where water is available for irrigation (Fig. 5–5). The dry, sunny climate appeals to some people for health, retirement, or other reasons. The human population of deserts will continue to increase as long as water is available.

TUNDRA

The **tundra** is a treeless, desolate expanse located between the water and ice of the Arctic Ocean and the continental timberline at the edge of the northern coniferous forest. In area it covers about ten percent of the earth's land surface. No similar large area is possible in the Southern Hemisphere because there are no comparable geographical conditions in Antarctica, though a few isolated islands have small areas of tundra. While the term *tundra* is the name of a habitat, it is also the name of a biome, a community with characteristic inhabitants.

Tundra forms a fringe around the Arctic Ocean along the top of North America and Eurasia. The largest areas of tundra are found in Canada and Siberia, but a considerable amount is located in Alaska. In addition to the Arctic location, an alpine form of tundra occurs on high mountains much farther south, where it occurs above the timberline.

The surface of the tundra is a rather featureless plain where former topographic irregularities have been smoothed out by Pleistocene glacial action. As a result of such action, the surface is scoured in places and coated with deposits in others. Long scratches and grooves in bedrock, scattered boulders, till, moraines, eskers, and a poorly developed drainage system remain as evidence of glacial activity. Because the drainage system is poor, most of the water simply collects and stands in low places during the short annual thaw. At that time about 50 percent of the surface is covered by water. This happens not only because of poor drainage but also because evaporation is low at such cold temperatures.

Freezing and thawing continue to change the face of the ground in unique ways. An example is the formation of geometric surface patterns. For miles the surface may be marked by circular or polygon-shaped patches varying in diameter from a few inches to over 300 feet. The patches may be mounds separated by trenches or shallow depressions with raised rocky rims (Fig. 5–6). The latter are probably caused by the expansion and contraction of different-textured materials as they freeze and thaw. Since finer materials like silt and clay contain more water, they expand farther as they freeze and push the coarser gravels and stones outward. Upon thawing, they shrink away from the heavy materials, leaving them where they were pushed. The endless freeze–thaw cycle creates a number of microhabitats as it sorts the particles and shapes the patches. Sometimes the rimmed depressions are good water-holding basins. On slopes, the circular or polygon-shaped patches may stretch into strips running up and down the incline. This is due to their slippage over **permafrost**, the permanently frozen subsoil. Besides causing "patterned ground," the action of ice sometimes breaks up bedrock into fields of angular boulders and heaves large volcanolike mounds (called pingos) as much as 300 feet above the surface.

FIGURE 5–6 *Patterned ground in alpine tundra along the Continental Divide near Estes Park, Colorado. Freezing and thawing are responsible for the polygon-shaped patches fringed by rocks.*

Precipitation is low, only about ten inches per year. Of that, about eight inches falls as rain during the summer season. The remainder is snowfall (equivalent to two inches of rain). In winter everything is frozen, but windswept highlands may be essentially free of accumulated snow. In warmer locations such a low precipitation rate would hardly support more than desert vegetation. In fact, the tundra is a "frozen desert" where water is locked in ice for much of the year and cannot be used. During the summer growing season, the upper few inches of soil thaw and become soggy, but the permafrost subsoil remains frozen. Even if other growing conditions were favorable, large plants could not grow without more root space. The soils and standing water are very acid.

Temperatures run the gamut of extremes from a low of −82°F (Canada) to a record high of 100° (Fort Yukon above the Arctic Circle). In Arctic regions the temperature fluctuates little in any twenty-four hour period owing to the long uninterrupted periods of daylight or darkness. Above the Arctic Circle, where the majority of the tundra is located, the sun shines continuously for weeks during the summer season, and immediately below the Circle the daylight hours are long. The winter season is long and harsh, marked by many successive days without sunlight or with very short days and long nights.

The tundra as a whole is relatively free from wind. For most of the nine-month winter it is a cold, dry, high-pressure area with little air movement. Winds are strongest when the seasons are changing.

Plant life consists of about 900 species, almost all less than four inches tall. It is especially interesting that about 400 of these species are found in both America and Eurasia. The plants usually grow in dense mats. Even the "trees" are dwarfed and sprawling. They consist primarily of willows, birches, and alders, all with trunks about as big around as a pencil and other parts in similar proportion. Such a stem may have 50 or more rings; larger willow stems about an inch in diameter have up to 400 rings. The ground-hugging habit of the plants keeps them close to the surface where it is warmer, often sheltered by rocks and other irregularities; here they benefit most from the reradiation of heat. Soggy places may be covered with sphagnum and dry ones with lichens. Between those extremes are habitats supporting seed plants like poppies, forget-me-nots, lupines, saxifrages, grasses, sedges, woodrush, buttercups, mustards, pinks, primroses, gentians, cranberries, blueberries, and heather. The ground cover may be sparse in unprotected, windswept places.

Animal life consists of both permanent and migratory residents, although a larger portion are migratory than in less severe climates. Mosquitoes, with unlimited breeding places, are abundant. The myriad pools and lakes, and a few streams, harbor lake trout, lake herring, whitefish, ling, smelt, northern pike, inconnu, arctic char, and grayling. In summer the aquatic habitats are thick with waterfowl. Some birds, like the ptarmigan and snowy owl, and mammals, like the arctic hare, collared lemming, and arctic fox, are as white as the background of snow on which they live. Others, though not white, tend to be lighter than their southern relatives. Some that are white in winter change into a darker coat when the snow melts. Many birds migrate into the tundra, where there is a good supply of insects and seeds for food. Better-known migratory birds include plovers and geese. Large herds of caribou (arctic deer) winter in the coniferous forest to the south and roam the tundra during the summer. There they are pestered by flies and mosquitoes and preyed upon by wolves. The musk ox is a tougher animal; it lives on the tundra throughout the year, as does the barren-ground grizzly. Not to be overlooked are the various lemmings that periodically reach high population levels and start migrating en masse. During the dreary winter months, many tundra animals seek shelter and food elsewhere or tunnel beneath the snow.

The endothermic animals in frigid climates must be able to conserve heat well in order to survive. They do this primarily with insulation: a dense coat of fur or feathers and a layer of fat located just beneath the skin. They can further conserve heat during inactive periods of sleep or rest by curling up into a ball. Strangely, many of these animals have widely varying temperatures between body and extremities. For example, the temperature in a reindeer's legs is sometimes 50°F lower than its body temperature.

Body size and shape are also involved to some extent in temperature regulation. When endothermic animals of the polar regions are contrasted with their counterparts in warmer climates in the low latitudes, some differences are sufficiently constant to be noticed. Although these observations were originally called rules, they have numerous exceptions. According to Bergmann's rule, animal bodies tend to be larger in colder climates. The significance of this adaptation is that polar mammals and birds can conserve heat better because large bodies have less surface area relative to internal mass. The opposite condition—small bodies with relatively more surface

FIGURE 5–7 *Alpine tundra near the Continental Divide in the Colorado Rockies. The ground-hugging mats of small plants give the alpine tundra an appearance much like that of Arctic tundra. The road in the foreground was used in the 1920s and has not yet been covered by vegetation.*

area—is equally useful to tropical animals that need to dissipate heat. According to another "rule," a trend known as Allen's rule, body extremities like snouts, ears, legs, and tails are shorter in colder climates and larger in warmer climates. Again, these differences can be related to the conservation or dissipation of heat.

Men native to the tundra have been hunters for generations. They are few in number and mostly nomadic. Outsiders entering their domain formerly fanned out over the barren wastes to trap the wild mammals for fur, but a new wave of humanity will soon sweep the tundra to take away its oil and mineral wealth.

Alpine tundra (Fig. 5–7) occurs on many high mountains outside the arctic region. It is located above the coniferous zone and below the snow cap (if present). In the Rockies tundra grows mostly between 11,000 and 14,000 feet, and in the Sierra Nevadas between 10,500 and 13,000 feet. In many respects alpine and arctic tundra grow under different environmental conditions; only the consequences as expressed in life forms are the same. Alpine tundra receives more wind, more precipitation, more cosmic and ultraviolet radiation, and more uniform exposure to light and darkness.

REFERENCES AND FURTHER READINGS

Dregne, H. E. 1970. *Arid Lands in Transition*. AAAS Publ. 90. Washington, D.C.

Hodge, C., ed. 1963. *Aridity and Man*. AAAS Publ. 74, Washington, D.C.

Jaeger, E. C. 1957. *The North American Deserts*. Stanford University Press, Stanford, Calif.

Ley, W. 1962. *The Poles*. Time-Life, New York.

McGinnies, W. G., B. J. Goldman, and Patricia Paylore, eds. 1968. *Deserts of the World*. University of Arizona Press, Tucson.

Wright, H. E., ed. 1968. *Arctic and Alpine Environments*. Indiana University Press, Bloomington.

Chapter 6
Streams

CHARACTERISTICS OF AQUATIC ENVIRONMENTS

With about 70 percent of the earth's surface covered by water, the potential exists for an unbelievable number of distinctive communities of organisms. No clear-cut or uniform system for classifying these communities is available. One logical way to handle the problem seems to be to recognize distinctive habitats and the organisms related to them.

Just because environments are aquatic does not mean they are uniform. In fact, temperature, depth, rate of movement, solute content, and solute concentration vary considerably. Add to that the selective nature of organisms to float, swim, attach, or burrow and niche possibilities become complex indeed.

Despite the variability of aquatic habitats, the water medium provides certain influences to which organisms respond in characteristic ways. For example, water absorbs or loses a large amount of heat without extreme changes in temperature. In this respect water moderates the environment. Except for a few **endothermic** (**homioi-thermic**, or warm-blooded) land animals that have returned to water to live or search for food, all aquatic organisms are **ectothermic** (**poikilothermic**, or cold-blooded); their temperatures approximate those of their surroundings. This eliminates the necessity for insulative coverings like hair or feathers. Surrounded by water, these organisms do not require waxy or oily surfaces to control evaporation. They may, however, have a slimy covering that minimizes friction as they move through the water. This is important, for water is about 780 times more dense than air and offers 100 times the resistance of air. Faster animals have streamlined bodies that further facilitate their movement. Because of the buoyancy of water, the weight-bearing skeletal systems of terrestrial animals and the woody structures of plants are either absent or poorly developed. Large numbers of aquatic organisms have simple reproductive systems that dump gametes into the water, leaving successful fertilization to chance.

Because of the solvent action of water, which dissolves more substances than any other liquid, various bodies of water differ markedly in the kinds and concentrations of their solutes. Just what is dissolved depends primarily on the contents of parent rocks and soils with which the water comes in contact. Depending primarily on the presence or absence of dissolved sodium chloride, ecologists separate aquatic habitats into freshwater and saline types. The distinction between the two is not precise, however. A contrast is often made between sea water, with about a 3.5 percent salt content (sodium chloride and other salts), and less salty (brackish or fresh) water. Some so-called freshwater streams are too salty for some types of life. For example, it is claimed that humans can use water that is unsuitable for horses, and camels can use water that is unfit for humans.

Analyses of fresh water used for human consumption show how much the solute content may vary from place to place. The differences are especially great between arid and humid locations. To illustrate, data from Yuma, Arizona, and Greenville, South Carolina, are given in Table 6–1. The water for Yuma comes from the Imperial Reservoir on the lower Colorado River, while that for Greenville

TABLE 6–1 *Comparison of mineral content of different water supplies*[a]

	Content in ppm	
Mineral	Yuma, Arizona	Greenville, S.C.
carbonate (as $CaCO_3$)	398	13
sulfate	367	1
chloride	129	4
calcium	100	2
magnesium	35	0.48
silica	13	2.50
iron	0.84	0.10

[a]Data from Greenville Water Works and the city of Yuma.

comes from reservoirs in nearby mountains where rainfall is high.

Light transmission through water is impeded and its quality altered just as light is altered when it passes through the atmosphere. The most obvious modifiers of light penetration are differences in wavelength energy, turbidity, and angle of incidence. Since most streams are shallow, the contrast between the quality of light at the top and at the bottom is not normally so extreme as it is in deep lakes and oceans.

Turbidity, or cloudiness that blocks light, may be due to loads of erosional sediments, plankton, or pollutants. Turbidity is more significant in streams and lakes than in oceans simply because small bodies of water do not have the volume to dilute the load as well. High water accompanying heavy rains invariably increases turbidity. The easiest way to measure turbidity is with a **Secchi disc**, a circular plate usually eight inches in diameter, which is lowered into the water by a chain (sometimes a cord or a rod) attached to its center. The disc may be entirely white or divided into four pie-shaped quadrants painted alternately white and black (or white and red). The disc is lowered into the water, and the depth is measured on the chain at the time when the disc is just about to pass from view. This measures the depth to which light penetrates and is reflected back to the eye and is related to the depth to which plants can survive. Incidentally, most aquatic plants seem to photosynthesize at their optimum levels at between 5 and 30 percent of full sunlight.

The angle of incidence (angle at which light strikes the surface) determines the amount of reflected light and the depth of penetration. Everything else being equal, midday sunlight from above penetrates deeper than sunlight from the rising or setting sun. Actually, the light travels the same distance through the water but not the same distance below the surface (Fig. 6–1).

Man has always used water as a dumping ground for his unwanted wastes. Today, harmful by-products from the agricultural and industrial enterprises necessary to support his exploding population endanger aquatic plants and animals.

Any study of aquatic communities of organisms has to take into account the changes wrought by both man and nature, the dearth of information about the vastness and depth of the oceans, and the extremes between fresh and saline bodies of water. A bubbling mountain brook has as little in common with the Great Salt Lake (with a salt concentration of up to 27 percent) as a farm pond has in common with the ocean. All have to be considered individually.

Within the aquatic domain, the many dissimilar inhabitants segregate themselves according to their mode of life. Consequently, they are designated **neuston** (on or hanging from the surface film), **plankton** (wafting about), **nekton** (swimming with progress), **pedon**

FIGURE 6–1 *The relationship of the angle of incidence of light to its depth of penetration in water. Compare beam 4, which is almost perpendicular, to beam 1, which intersects the surface at an oblique angle. (From C. Leland Rodgers,* Essentials of Biology, *©1967, Barron's Educational Series, Inc., Woodbury, New York.)*

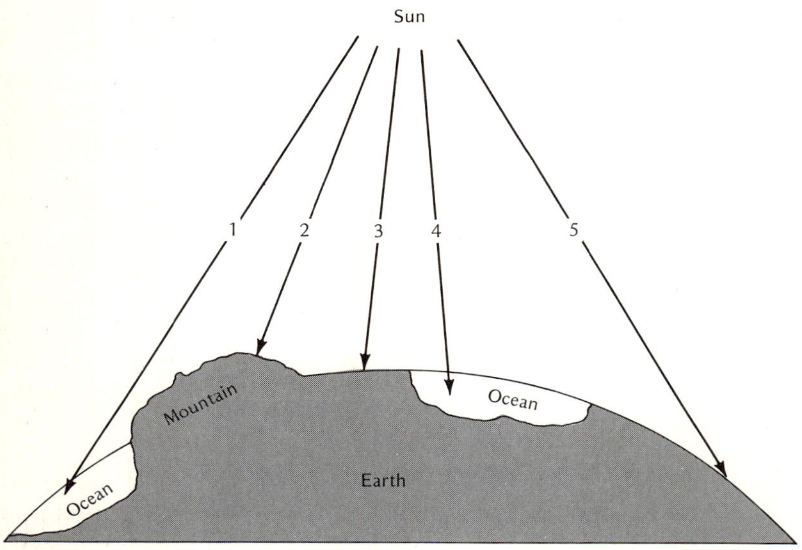

or **benthos** (on or in the bottom), and **periphyton** (attached to plants or other objects).

LIMNETIC (FRESHWATER) HABITATS AND COMMUNITIES

Probably no more than 0.002 percent of the earth's total supply of water is flowing or impounded on the continents. For convenience, the freshwater habitats are separated into two fundamental groups —**lotic** and **lentic**. The lotic habitats are flowing streams, whereas the lentic ones are bodies of standing water like ponds and lakes. That difference alone is enough to account for some basic differences in their communities.

LOTIC HABITATS (STREAMS) AND COMMUNITIES

The major differences among streams are a matter of volume of water and speed of flow. Even a single stream varies in these two respects so that one segment of that stream is often more like another stream than other segments of the same stream. Although there is no ideal way of classifying streams into distinctive groups, the geologists' way of distinguishing them, or segments of them, by "age" is useful. By age geologists mean erosional (or depositional) stages, not fixed periods of time. The stages are **youth**, **maturity**, and **old age**. Each is directly related to the channel gradient and hence to speed of flow, which in turn is determined by the gradient.

A **youthful stream** (Fig. 6–2) is typical of mountainous areas. On the steep slopes water flows rapidly and erodes proportionally. Streams transport solids by carrying smaller particles in suspension and rolling or pushing larger pieces along the bottom. The faster the water flows, the larger the particles it can move. Heavy rains in mountain valleys are particularly destructive because the runoff is confined to narrow, steep channels. Large boulders weighing many tons are sometimes moved along by such force. Naturally, the finer particles are carried completely away. The speed of flow accounts for the scarcity of mud or fine sand in mountain streams and for the preponderance of coarse sand, gravel, and stones.

The abrasive effect of the transported material against the bottom rapidly deepens the channel so that the stream bed is at the bottom of a V-shaped valley as seen in cross section. The harder rocks in the stream path wear more slowly and result in the formation of rapids and riffles. Youthful streams tend to flow straighter than older streams as they respond to steep gradients, unencumbered by the clogging deposits that occur in more sluggish water. At this stage there is no floodplain. The water in youthful streams is shallow except for small pools such as those found below waterfalls.

Inhabitants of youthful streams have to contend with rapid water. This they do by anchoring themselves, avoiding the current, or swimming vigorously. Some of the organisms, especially algae, are slippery and offer little resistance to the flow of water. Both plants and animals attach themselves to stones, congregating on the side away from the full force of the flow.

The common plants inhabiting youthful streams are algae and mosses, although a seed plant known as riverweed (*Podostemum ceratophyllum*) grows profusely on rocks in some eastern American streams. This plant looks like some seaweeds and is easily mistaken for an alga. Matted plants of all types provide shelter for small organisms or larvae like protozoa, diatoms, rotifers, tardigrades, caddis fly larvae, and mites.

Animals are often anchored by hooks (immature blackfly) or suckers (midge larva), or hide under plants, stones (mayfly nymph),

FIGURE 6–2 *A youthful stream in the Appalachian Mountains. Such streams are characterized by waterfalls and rapids.*

or debris. Some have body shapes that offer minimal resistance to flowing water; examples are the very flat, mound-shaped water-penny and the larva of the psephenid beetle. The adult beetle, on the other hand, has clawed feet for clinging. Particularly common in swift-flowing streams is the caddis fly larva, which cements around itself a protective cocoon composed of sand grains, tiny bits of stone, and other materials. Some of the larvae construct nets to capture food. The caddis fly, stone fly, and hellgrammite (larva of the dobson fly) live primarily under rocks where they are sheltered from the currents. Blackfly larvae (*Simulium*), attached to the downstream side of stones away from the full force of the currents, are at times so abundant that they are called "black moss."

Water in youthful streams is well oxygenated by the agitation of passing over rapids and falls. Salamanders generally indicate this condition by having smaller lungs (adults) or gills (larvae). The mayfly nymphs living in the highly oxygenated water of swift streams have gill surfaces up to two-thirds smaller than those of closely related species living in mud-bottomed pools where oxygen is limited. It is likely that many other animals also have reduced respiratory organs. Mayfly nymphs living in pools where oxygen is less abundant have the peculiar habit of flicking their gills back and forth, some all the time and others only when oxygen is scarce. This is an unnecessary habit for animals living in moving water, which is usually rich in dissolved oxygen. In some animals there is evidence that the rate of metabolism is related to oxygen supply, which means that in oxygenated streams they can either be more active or maintain the same rate of activity with smaller respiratory organs.

The water in mountain streams is colder than in older streams because of higher altitudes and greater shading. Temperatures remain fairly constant. At a fish hatchery in Tennessee where rainbow trout are reared, the temperature is approximately 56°F all year. The temperature drops to freezing at times, but solid ice generally forms only along the edges and not where the water is swiftly moving. The moving water may, however, become slushy with ice crystals. Mountain trout are accustomed to cold water and cannot survive at high temperatures. Rainbow and brook trout thrive in streams where the summer temperature does not exceed 70°F but cannot survive even short periods of exposure to temperatures much above 75°F.

Among the irregularities of a cascading stream are places where the water is comparatively still and deep. To some animals these pools are more appealing than turbulent water. Such pools contain crayfishes, salamanders, and various more mobile fishes like trout, darters, dace, and suckers. Fishermen go for the trout because of their size and gameness. This gamy fish feeds on aquatic insects, adult and immature, and an infinite number of other insects that fall from the trees and bushes overhead. Nymphs of mayflies, damsel flies, and dragonflies are common in the pools. Top-dwellers supported on the surface film of the water include whirligig beetles and the spidery water striders (a type of bug). Other occasional inhabitants of the pools are the invertebrate planarians and snails, and the vertebrate frogs, snakes, turtles, and otters.

The characteristics of **mature** and **old streams** are almost entirely different from those of youthful streams. The water flows gently, contains less oxygen, is warmer, and is more turbid. The bottom is composed of sand, silt, or mud, generally intermixed or covered with a considerable amount of decaying organic matter. Often marginal vegetation, and sometimes floating vegetation as well, adds considerably to the food supply and organic accumulation. Communities supported by such streams are more like those found in lakes than those of mountain streams.

The bottoms, which are rich with decaying matter, attract nematodes, a number of immature insects, and bottom-feeding fishes like cat, carp, buffalo, and sucker. Other fishes found in mature streams include minnows, drums, sunfishes, crappies, and bass. Also present are salamanders and frog tadpoles. Noteworthy are the large numbers of mosquito and midge larvae, and damsel fly and dragonfly nymphs; these provide food for some of the fishes. Some sheltered marginal pools and water among the rooted aquatics may be sufficiently calm to develop a good crop of plankton, also an important food source. Plankton is composed of small, mature or immature organisms living near the surface and drifting with water currents. Also found in some areas are large populations of sponges, hydras, bryozoans, leeches, clams, and snails.

Mature streams are those that have reached grade, a point in the erosion cycle when downcutting eliminates most falls and rapids, giving the stream bed a gentle slope (Fig. 6–3). Lateral erosion then replaces vertical erosion as the dominant process sculpturing the valley. Two changes are in progress—the valley floor is widening into a floodplain and the steepness of the valley slopes is diminishing. The stream's course becomes more serpentine. The flowing water moves faster on the outside of the curves; it has a corkscrew-like motion, sweeping across the bottom from outside to inside and back again across the top to the outside. This peculiar motion results from unequal flow caused by centrifugal force and frictional drag on the bottom and sides of the channel. The combined flowing action erodes the outer banks and bottom while building gravel, sand, and mud bars on the shorter inner side of the curves. The net effect is that the curves become larger and larger.

Animals living either on the surface or in and out of the water are numerous. Whirligig beetles and water striders are ubiquitous where the water is still. Frogs, turtles, snakes, otters, muskrats, and minks

FIGURE 6–3 *A mature stream, the Holston River near Kingsport, Tennessee. The stream bed has a gentle slope and is free of falls and rapids. Note the fringe of bottom-land trees along its banks. They are sycamore* (Platanus occidentalis), *willow* (Salix nigra), *box elder* (Acer negundo), *and ash* (Fraxinus).

are typical amphibious animals. Alligators and crocodiles are more restricted geographically.

Because old streams are close to sea level, they have little gradient and a sluggish flow. Such streams transport only finer materials like sand, silt, and clay, which they deposit on the floodplain or even in the channel itself. The clogged channels force the streams to relocate along freer routes. Clogging interferes with drainage, and water left behind after rechanneling stands in sections of the old river bed. Thence, the wide valley floor is replete with swamps and lakes. The stream meanders in wide sweeps forming oxbow curves. Eventually it takes a shortcut bypassing the curve, which remains as an oxbow lake.

The difference between old streams and mature streams is a matter of degree. Now close to sea level, old streams have swampy connections, backwaters, and lakes, all filled with organisms like those in mature streams. In the estuaries there is, of course, some mingling of freshwater and marine organisms.

HOT SPRINGS

Extreme environmental conditions, like the high temperature of hot springs, are generally unfavorable to life. Studies have shown, however, that for organisms adapted to hot springs the high temperatures are optimal. This means that these organisms have been selected through evolution for such habitats and are not there simply because of reduced competition.

Hot springs, like other aquatic habitats, run the gamut of environmental variations. Flowing as they do through all types of rock formations, they dissolve and transport many kinds of solutes—pH values range from 1 to 9. Besides the chemical differences, temperatures of hot (or warm) springs range from close to groundwater temperatures to boiling. At Hot Springs, Arkansas, the average temperature of the springs is 143°F, and at Yellowstone many springs are boiling (Fig. 6–4). Although temperature and other conditions vary from one spring to another, each one has a steady, distinctive year-round temperature.

Good comparisons of the suitability of life to different springs are impossible to make because of the many environmental variables affecting them. However, a single spring affords an exceptional opportunity to compare growth rates of organisms in water with the same chemistry but at different temperatures. Organisms can be

FIGURE 6–4 *A boiling hot spring (Morning Glory Pool in Yellowstone National Park).*

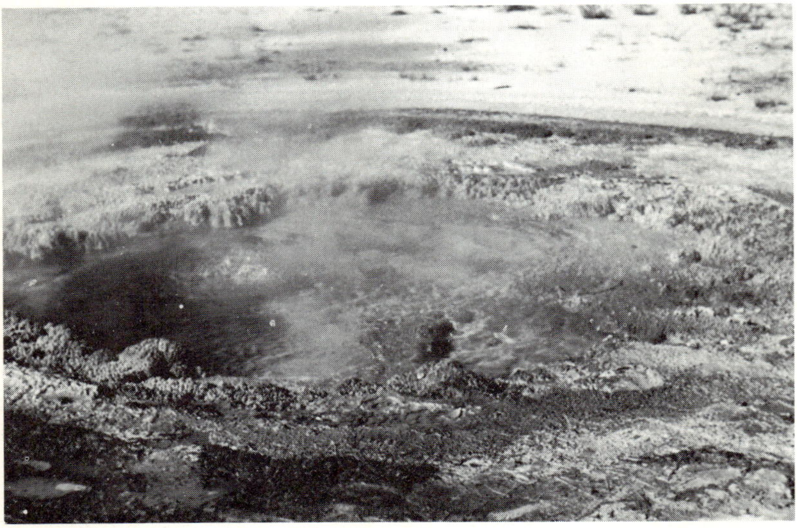

studied along a temperature gradient from the outlet of the spring to cooler points downstream.

Only a few plants and animals can live at temperatures much above 100°F, yet studies made of life in Yellowstone springs show that high temperatures are not adverse to some organisms, often referred to as **thermophiles**. Some bacteria actually thrive in boiling water, but photosynthetic plants cannot live at such high temperatures. Procaryotic blue–green algae are found at temperatures up to 162–167°F, and certain fungi live in temperatures up to 132–140°F. Animals are less tolerant of heat, but some protozoa have upper limits of 113–124°F.

SUBTERRANEAN STREAMS AND POOLS

The cave environment, with its uniform darkness, humidity, and temperature, contrasts sharply with the outside environment. The total or near absence of light virtually eliminates photosynthetic activity, visual responses, and circadian rhythms, all of which are light-dependent. Humidity is high, usually between 95 and 100 percent. The temperature approximates the mean outside temperature for the region. Although the temperature of a specific cave fluctuates little, those of caves at different altitudes or latitudes may vary by many degrees. Some may be sultry, others perennially icy.

Within the cave environment are two aqueous habitats: closed pools fed by water dripping from above and streams that course through fissures and galleries. Stream systems are often inaccessible and untraceable except for short distances. Sometimes well drillers strike the channel and bring up a surprising number of specimens from deep below the surface. A few years ago several dozen salamanders were collected in this manner in a Karst region of east Tennessee. The underground streams have the usual stream features such as bars, falls, and pools. They may or may not be freely connected with outside streams. Cave streams connected with the outside are better places for life because they have a lower pH and more organic matter available for animal food. The closed pools have a high pH due to the load of dissolved carbonates in the dripping water that feeds them. With little organic matter in the water, there is not enough food to support more than a sparse fauna.

Caves may be divided into a **twilight zone**, a **middle zone**, and an **interior zone**. The twilight zone is near the entrance, where some light is available and there are a considerable number of animals moving between the cave and the outside environment. The light is sufficient for some photosynthetic plants. The middle zone is an area of complete darkness and variable temperature. The interior zone has complete darkness and a constant temperature. It displays a combination of conditions that represent an extreme variation from the outside environment.

Life in the pools and streams of caves, particularly in the interior zone, exhibits responses more or less keyed to uniform temperatures, absence of light, absence of a day–night cycle, lack or near lack of predators and pathogens, reduced competition, and scarcity of food. Under evolutionary pressure, fishes have responded to varying degrees through degeneration of the eyes and the visual centers of the brain, increase in sensitivity of the lateral line system, degeneration of pigment cells, loss of circadian activity, loss of escape responses, reduction in rates of growth, lowering of metabolic rates, and increase in length of life.

Organisms typical of the pools and streams in the interior zone are bacteria, amphipods, isopods, crayfish, salamanders, and fishes. Some of the inhabitants have gone to the evolutionary extreme in response to the unique environment and become obligate residents, whereas others are only facultative. Animals sometimes enter caves as regular or accidental visitors.

REFERENCES AND FURTHER READINGS

Brock, T. D. 1967. "Life at High Temperatures." *Science 158:* 1012–1019.

Coker, R. E. 1968. *Streams and Ponds.* Harper & Row, New York.

International Symposium on Eutrophication. 1969. *Eutrophication: Causes, Consequences, Correctives.* National Academy of Sciences, Washington, D.C.

Morgan, Ann H. 1930. *Field Book of Ponds and Streams.* Putnam, New York.

Poulson, T. L., and W. B. White. 1969. "The Cave Environment." *Science 165:* 971–981.

Welch, P. S. 1952. *Limnology.* 2d ed. McGraw-Hill, New York.

Chapter 7
Lakes

LENTIC HABITATS
(LAKES, PONDS, SWAMPS, MARSHES, BOGS)

Unlike lotic water, which flows steadily and continuously, lentic water stands in basins that may or may not have outlets. The movement of some water toward the outlet, if there is one, and that of waves on the surface do not alter the basic characteristics of an impounded body of water.

Mention has already been made of the relationship between the movement of water and its ability to transport and erode. As soon as speed of the water slackens, sediments begin to settle to the bottom. Bodies of still water are therefore accumulators of both sediments brought by feeder streams and organic remains from organisms indigenous to the body of water itself. Allowed enough time, the basin will fill.

The distinctions among the different lentic environments are not always perfectly clear. In general, **lakes** are characterized as large bodies of open water; **ponds**, as small bodies of open water; **swamps**, as waterlogged soil or shallow water with trees; **marshes**, as waterlogged soil or shallow water with emergent vegetation but treeless; and **bogs**, as marshes often vegetated with sphagnum (Fig. 7–1) and heaths. Swamps, marshes, and bogs all have above-the-surface vegetation and hold standing water part of the time. The water may be a few inches to two or three feet deep and may persist for months at a time. On the other hand, it may disappear or remain only in depressions during prolonged dry periods. As used here, swamps, marshes, and bogs may be separate entities or zones marginal to streams or lakes.

FIGURE 7–1 *Sphagnum moss in the Okefenokee Swamp.*

LAKES

Lakes are large inland bodies of open water. The implication is that they may be large enough to have some wave action, although in smaller lakes this would not amount to much. Also, the water is usually deep, at least at some point in the basin.

Lakes may be both natural or man-made. In either case, water settles in low places or is entrapped by blocked streams. Lakes abound in glaciated regions, where the relocation of crustal materials by scouring and deposition has completely destroyed the established drainage system and a new one has not had enough time to develop. Other natural causes of lakes include sinks in limestone country, landslides, volcanic activity, diastrophism, and sedimentation. The latter causes a stream to become clogged and change course, leaving a lake in the abandoned part of the channel. Artificial lakes (Fig. 7–2) are built primarily for municipal water supplies, irrigation, flood control, generation of electricity, improvement of navigation, or recreation.

The surface area and depth of lakes vary to such an extent that their physical, chemical, and biological characteristics are often highly individual. The world's largest freshwater lake is Lake Superior, with an area of more than 31,000 square miles and a max-

FIGURE 7–2 *A reservoir serving a city in Chihuahua, Mexico.*

imum depth of about 1290 feet. The deepest is Lake Baikal in southern Siberia, which is over a mile deep (5600 feet). Of volcanic origin, the spectacular Crater Lake in Oregon is about 2000 feet deep.

Because of the extreme differences among lakes, they are conveniently contrasted as **freshwater** or **saline**, **shallow** or **deep**, **warm** or **cold**, and **large** or **small**. Obviously, such categories are inexact, even as used by ecologists.

In addition, the large lakes may be divided into strata and zones as follows:

I. Strata (one-above-the-other positions)
- **A.** Based on temperature (Fig. 7–3)
  1. **Epilimnion**—warm layer, gradual drop in temperature from top to bottom
  2. **Metalimnion** (mesolimnion or thermocline)—sharp drop in temperature from top to bottom, about 1°F decrease per two feet in depth
  3. **Hypolimnion** (profundal stratum)—sharp drop in temperature from top to bottom
- **B.** Based on predominant metabolic activity
  1. **Trophogenic stratum**—photosynthesis (building-up process) dominates

FIGURE 7–3 *Lake stratification and zonation.*

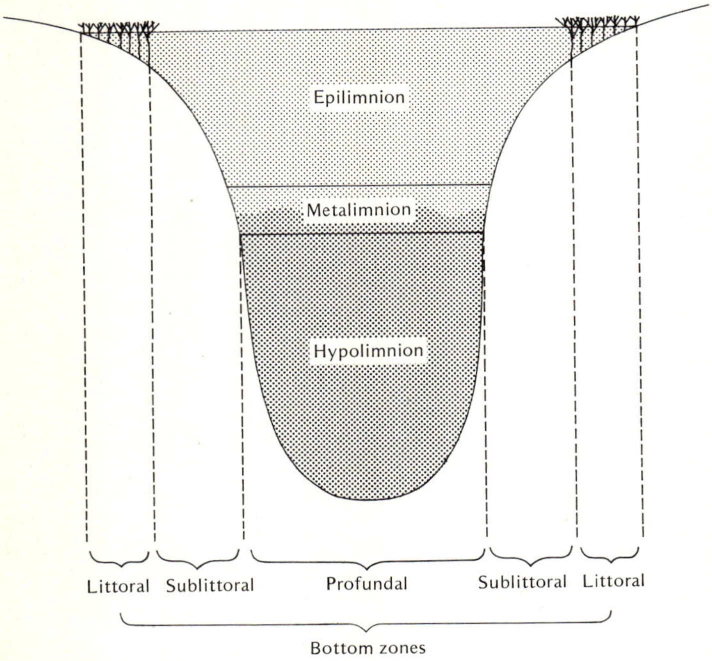

2. **Compensation level**—boundary between trophogenic and tropholytic strata where oxygen production from photosynthesis is equal to oxygen consumption in respiration
3. **Tropholytic stratum**—respiration (tearing-down process) dominates

C. Based on light penetration (about the same as in B)

1. **Photic stratum** (zone)—with effective light
2. **Aphotic stratum** (zone)—without effective light

D. Based on location

1. **Surface**—part of lake covered by atmosphere
2. **Limnetic stratum** (zone)—all open water above bottom (divided by some into an upper limnetic stratum and a lower profundal stratum)
3. **Bottom** (benthos)—part of basin covered by water

II. Zones of lake bottom (side-by-side position, Fig. 7–7)—limnologists do not agree on the limits of these zones

A. Littoral zone—under shallow water, often with rooted vegetation

B. Sublittoral zone—where epilimnion and metalimnion intersect the bottom; from rooted vegetation to where the hypolimnion intersects the bottom

C. **Profundal zone**—essentially underneath the hypolimnion (including, according to some authors, the stratum of water above the profundal zone)

PRODUCTIVITY OF LAKES

Lakes may be classified according to productivity, the rate of increase in biomass based on photosynthesis. Among the factors influencing the rate of photosynthesis are available carbon dioxide, temperature, nutrients, and light. One or more of these factors may differ sharply in any body of water, and photosynthesis is affected accordingly.

In any deep body of water like a lake or an ocean, both quality and quantity of light available for photosynthesis at any particular depth is related to how much and how far that light can penetrate the water. Even if the light is not blocked by turbidity, there are limits to which it can penetrate. Both the angle of incidence (see Chapter 9) and the wavelength of light determine its availability and effectiveness at different levels.

Visible light penetrates to little more than 300 feet in sufficient quantity to supply energy for photosynthesis. That depth marks the lower limit of what is known as the **photic stratum**. At the bottom of this stratum, the quantity of carbon dioxide released by respiration and that of oxygen released by photosynthesis are substantially balanced (**compensation level**). The long thermal rays below the visible spectrum are absorbed close to the surface, within the upper few inches. Wavelengths in the blue portion of the spectrum (475 nm) penetrate to the greatest depth. In the clearest water only about 1 percent of the blue light reaches a depth of 435 feet. All of these figures are close to the maximum depths possible under ideal conditions, which are seldom achieved except during short periods.

Related to productivity is the amount of oxygen dissolved in the **hypolimnion**, or lower stratum of a deep lake, during the growing season. This is the time of the year when most organic matter is produced, and it naturally follows that it is also the period when a great deal of organic residue settles to the bottom. Decay organisms act on the organic matter and consume oxygen in the process. Consequently, oxygen is depleted in proportion to the amount of organic matter being decayed by the bacteria and other decomposers. Low oxygen content in the hypolimnion, therefore, can be an indication of high productivity by photosynthetic plants above. Measured by the standard of productivity as indicated by the oxygen supply in the hypolimnion, lakes may be designated **oligotrophic**, **eutrophic**, or **dystrophic**.

The oligotrophic lake (Fig. 7–4) is usually more than 55 feet deep. It contains little organic matter because only a smal! portion of its water is suitably lighted for photosynthesis. Its basin has such steep sides that its littoral zone is either lacking or limited to a narrow fringe. The hypolimnion is relatively rich in oxygen because of the dearth of organic matter and the low level of decay. The fish population consists of such species as lake trout, whitefish, and cisco. Allowed enough time, the lake will evolve into a eutrophic lake as it fills with sediments.

The eutrophic lake (Fig. 7–5) is by far the most productive, as indicated by an abundance of organic matter and nutrients. It is shallow enough (under fifty-five feet) for a large portion of its water to be lighted and suitable for autotrophic plants. It normally has well-developed littoral flora and an abundance of **phytoplankton** capable of supporting a rich fauna. An abundance of decaying organic matter on the bottom depletes the oxygen supply in the hypolimnion during the growing seasons. Seasonal turnovers periodically renew the oxygen supply from above and the carbon dioxide and mineral nutrient supply from below (released by decay). The carbon dioxide

FIGURE 7–4 *Crater Lake in Oregon, an example of an oligotrophic lake.*

FIGURE 7–5 *Eutrophic lake with several zones of vegetation. The plants with floating leaves are water-shield* (Brasenia schreberi)*; the larger aquatic plants in the background are spatterdock (*Nuphar luteum*). A luxurient growth of water-weed* (Elodea canadensis) *and bladderwort* (Utricularia biflora) *is submerged in the open-water zone.*

and nutrients then stimulate rapid plant growth. Fishes like largemouth bass, pike, sunfish, and perch live in the epilimnion, or upper stratum.

Many eutrophic lakes are too shallow to show temperature stratification. In essence, a shallow lake is like the epilimnion of a deep lake, rich with plants and regularly circulated. Allowed enough time, the eutrophic lake will fill with sediments and evolve into a shallow vegetated place such as a marsh.

The term **eutrophication**, from the same root word as eutrophic, is currently popular among people talking about water pollution in either lakes or streams. Pollutants often increase productivity by stimulating the growth of large populations of algae and water weeds that upset natural balances. A case in point would be fish kills resulting from the disturbance of the oxygen–carbon dioxide ratio. One cause of explosive algal growth is enrichment with phosphorus contained in detergents and other materials that are dumped as waste into the water.

The dystrophic lake is distinctive because climatic conditions make it unproductive. It is a cold-climate lake with a great deal of marginal vegetation and organic accumulation. What is missing is the complete decay of vegetable matter. The cold climate and anaerobic conditions together result in partial decomposition of organic matter to peat and organic acids. Because nutrients like nitrogen, phosphorus, and potassium are still locked in the peat, the water is infertile. Because phytoplankton is scarce and photosynthesis is limited, the oxygen content is low and the carbon dioxide content high. Fishes are usually scarce. In time the dystrophic lake will develop into a bog.

MOVEMENT OF WATER IN LAKES

In any lake there are different kinds of water movement, depending on area and depth. To what extent a particular kind of motion may affect the flora and fauna depends on such things as magnitude, time, and frequency. Tides are imperceptible except in large lakes, and even there they are too small to be of any biological consequence (about two inches in Lake Michigan). Wave action is unimportant in small lakes but batters the littoral zone in large ones. Two other kinds of motion, **turnover** and **seiche** (sash), are of greater importance.

Motion in a lake is strongly related to the size of the lake. Since wind is the primary cause of motion, the large lake is most affected because it exposes a greater surface to wind. Depth is also important in influencing turnover.

A lake less than thirty to thirty-five feet deep is too shallow to be thermally stratified and is thoroughly mixed by constant turnover except when the surface is frozen. Wind blowing on its surface simply pushes water forward, piling it on the opposite shore. After reaching the limit of its forward motion, the water responds to the stronger pull of gravity and reverses direction, then follows the course of least resistance underneath the surface current. No complications arise because all the water is of the same density.

The story is different in deep lakes in which strata of different densities develop. The temperature of the upper water fluctuates with the temperature of the air above it, but that of the deeper water, which is insulated from the air, remains unaffected. This difference in temperature causes corresponding differences in the density of the two layers, a condition that interferes with their mixing. They can mix freely only during short seasons of the year when they are at approximately the same temperature; then they are similar enough in density for the wind to circulate the entire lake.

In temperate regions turnover (Fig. 7–6) occurs twice yearly, in

FIGURE 7–6 *Turnover in lakes. (a) Shallow lake at any season except when surface is frozen. (b) Deep lake in summer. (c) Deep lake at time of spring and fall turnover. (d) A lake of any size frozen over in winter; there is no circulation of water.*

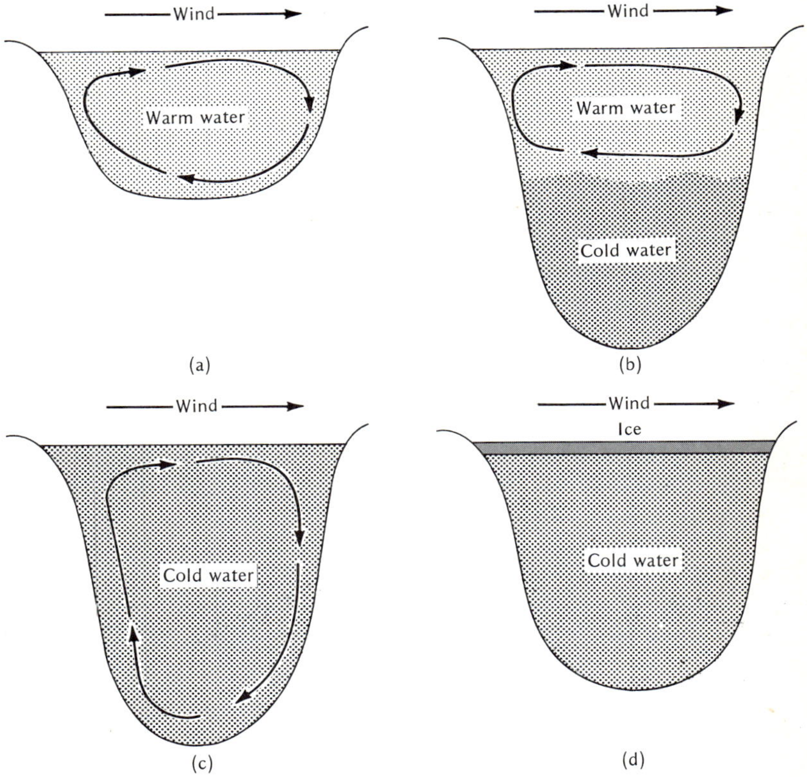

spring and autumn. During the summer the warm epilimnion overlies colder and denser water. The coldest and densest layer is the hypolimnion. Separating the two is a stratum of water in which there is a rapid transition in temperature that amounts to a decrease of about 1°F per two feet in depth. The transition stratum is the metalimnion, often called the **thermocline**. In summer the epilimnion circulates over the hypolimnion the same way water circulates in a shallow lake.

With the passing of summer, the epilimnion begins to cool and gradually expands downward until it merges with water of the same temperature located deeper in the lake. Thus, uniformity of temperature is achieved throughout the lake when the water cools to about

39°F, the point of greatest density. Wind blowing against the surface can then circulate the entire lake the same way a shallow lake is circulated. This is the period of autumn turnover and it may last for several weeks.

As the temperature drops below 39°F at the surface, the turnover ceases because the upper water expands into a lighter layer that floats on the heavier lower water, which is still around 39°. This condition, during which colder water overlies warmer water, is referred to as a **temperature inversion**. If the temperature drops low enough, the surface freezes and the water is protected from the wind until the ice thaws. Upon thawing, the water warms to approximately 39°, at which point the temperature is again uniform throughout the lake. Wind blowing against the surface sets it in motion in the **spring turnover**, which may also last several weeks. It is terminated when the surface warms and an epilimnion is reestablished.

Turnovers are likewise possible in deep lakes at temperatures above 39°F. They occur when the temperature of the epilimnion falls close enough to that of the hypolimnion so that wind-driven currents can overcome the density differential and mix the water. When the temperature rises again, the density differential reestablishes stratification.

Turnovers are important to keep gases and nutrients in balance throughout the lake. Each summer, photosynthetic plants increase the dissolved oxygen concentration and deplete the carbon dioxide concentration and certain nutrients in the trophogenic stratum. At the same time, respiring organisms, including the decomposing organisms, deplete the oxygen content and increase the carbon dioxide levels in the tropholytic stratum. In addition, decomposition releases nutrients from dead bodies. If periodic mixing did not occur, the lake would soon be devoid of these recycled substances and become quite unproductive.

A seiche of a large lake is like water splashing in a pan. The water first piles up on one side and then oscillates back to the other. Once instigated, the oscillation continues with diminishing amplitude until frictional drag on the lake basin or some other counteracting force stops it. The rocking motion begins with a temporary, localized elevation or depression caused by strong wind, sudden change in atmospheric pressure, heavy rainfall, earthquake, or landslide. Doubtless the most common cause is wind. Except in large lakes, seiches are too small to be noticed. On large lakes, however, they may reach a height or depth of several feet.

Unseen internal seiches occur during the summer in lakes that are thermally stratified. They may be detected by taking temperature readings at different depths at one location over a period of time. At the point of measurement one may find that temperature levels are not constant but move up and down. The oscillation occurs between the epilimnion and the hypolimnion as the two layers flow in opposite directions (Fig. 7–7). For example, steady wind removes surface water from one shore and piles it up on the opposite shore and in so doing tilts the hypolimnion in the opposite direction. Needless to say, this creates turbulence that produces a considerable amount of mixing in both layers. For this reason, internal seiches are of greater biological importance than those at the surface.

FIGURE 7–7 *One type of seiche. (a) Surface water blown by wind piles up on Shore B. This results in downward displacement of the hypolimnion at that side of the lake. (b) Oscillation begins as wind ceases. The water currents reverse direction as the hypolimnion readjusts in response to gravity; the surface water begins swinging back to Shore A. (c) First oscillation completed as full force of surface water is spent on Shore A. This is followed by oscillations of diminishing intensity until the water is calm and the metalimnion is level.*

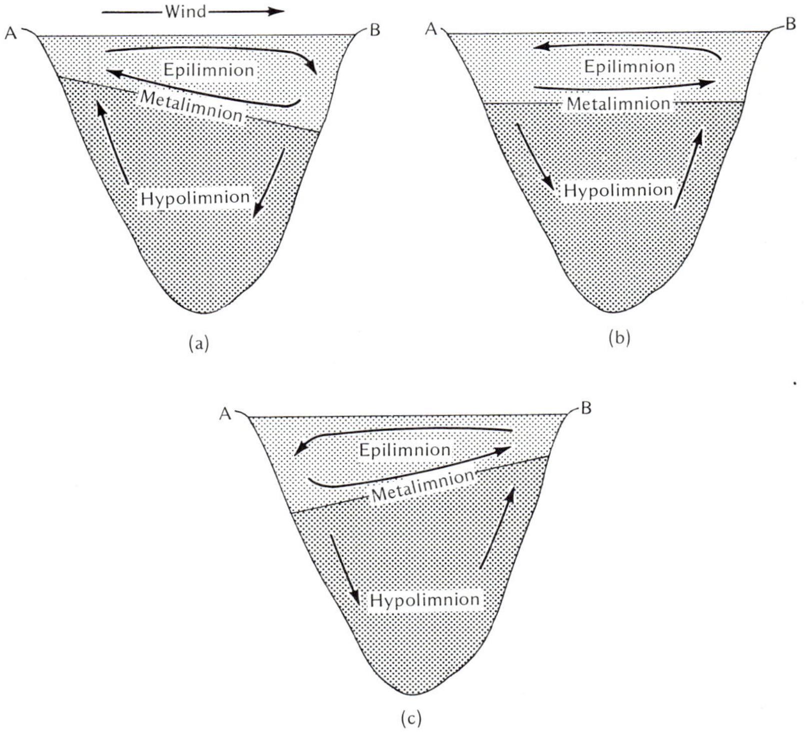

LIFE IN LAKES

The **neuston** consists of organisms associated with the surface film, either supported on it or hanging beneath it. It includes both plants and animals. Representative plants supported on the film are some species of desmids, bacteria and other fungi, water ferns (*Azolla* and *Salvinia*), and seed plants of the family Lemnaceae, such as duckdeed (*Lemna*) and water-meal (*Wolffia*). Animals living on the surface film include whirligig beetles, water striders, spiders, and spring-tails. There are also clusters of mosquito eggs. Attached to the underside of the surface film are hydras, flatworms, small snails, cladocerans, ostracods, wigglers (mosquito larvae), and the larvae of some midges. The surface film, while useful to the organisms mentioned, may act as a barrier of sorts to weak insects like mayflies attempting to pass from the water into the air.

Plankton grows well and accumulates in standing bodies of water because it is not washed away. Generally, a considerable number of species may be found at any particular time, although their relative abundance can vary widely within a short period. It is a well-known fact that a single species will occasionally multiply at a very rapid rate when conditions are ideal, causing the population to bloom conspicuously. **Blooms**, or large populations resulting from high productivity, result primarily from enrichment (eutrophication), often following fertilization or pollution.

Whether enrichment is desirable in either lakes or streams depends on how it affects natural balances or how the water may be used. It is desirable to fertilize fish ponds and lakes to increase the food supply, although one risks upsetting the oxygen balance. Overfertilization can cause so much plant growth that photosynthesis will raise the oxygen level high enough to kill fishes. Fishes also die when oxygen is deficient, as when there is too little photosynthesis, or when oxygen is depleted by decay organisms. Blooms may become serious when a reservoir or stream used for drinking water has a population surge of foul-tasting or -smelling organisms that the treatment system cannot cope with.

Eutrophication may occur in environments other than lakes, such as oceans. The Red Sea, for example, is named for the blooms caused by the blue–green alga *Trichodesmium*. Some have extended the application of the term to land environments, where evidence indicates that air polluted with nitrate and sulfate accelerates the production of biomass.

Where plankton occurs, it is likely to include diatoms; miscellaneous green algae; bacteria and other fungi; protozoa; small crustaceans like copepods, ostracods, and cladocerans; rotifers; and reproductive cells and immature stages of a variety of organisms.

Less frequent but often locally abundant are blue–green algae, water mites, gastrotrichs, and freshwater jellyfish (*Craspedacusta*). That many planktonic species are cosmopolitan in distribution is not surprising, considering the ease with which they can be transported by waterfowl.

The nekton, or swimming animals, consists primarily of fishes. The common species are bass, sunfish, crappie, warmouth, rock bass, bluegill, lake trout, pike, carp, sucker, buffalo, walleye, landlocked salmon, landlocked smelt, muskellunge, pickerel, black-nosed dace, perch, bullhead and other catfish, burbot, freshwater drumfish, and lake whitefish.

According to their nature, fishes select depths and temperatures that suit them. Fishes in deep lakes, therefore, often change levels at different seasons. This is illustrated by the positions of a few common fishes in lakes of temperate climates during the summer and winter. In summer, bluegills, perch, and bass are active feeders in the warm epilimnion. When winter comes and the water temperature is nearly uniform, bluegills and bass sink lower and become sluggish; the perch remain active and expand their range to include all depths. Whitefish remain at the thermocline in summer and at approximately the same level in winter even after the thermocline has been obliterated. Lake trout select the cold bottom of the hypolimnion during the summer and expand their range throughout the lake in the winter as soon as all the water gets cold. The bottom-feeding catfishes stay in shallow water along the banks. They are active in summer but hibernate in the mud during the winter.

Other than the fishes, most nektonic animals are actually part-time nektonic, either moving in and out of the water or inhabiting the water only at some stage of development. Most of them are insects or vertebrates. The better-known nektonic insects are back swimmers, (true bugs), water boatmen, (bugs), water scorpions, (bugs), giant water bugs (also called electric-light bugs), midge larvae (also called phantom larvae), mosquito larvae, predaceous diving beetles, and water scavenger beetles. Important nektonic vertebrates are salamanders, frogs, turtles, snakes, muskrats, otters, and minks.

The pedon (benthos), life on or in the bottom, is distributed in a variety of habitats that extend from the shallow edge to the greatest depth. Of the habitat variables, the composition of the bottom may range through all textures from rocky to muddy. The littoral zone, where the water is shallow, is normally the home of a variety of rooted plants. This zone is especially rich in insects and molluscs but also contains sponges, bryozoans, flatworms, nematodes, and annelids.

Because of the usual scarcity of food and oxygen, populations are relatively sparse in the profundal zone. Profundal organisms are small, mostly microscopic. They include protozoans, rotifers, nematodes, gastrotrichs, cladocerans, copepods, ostracods, tradigrades, insect larvae, water mites, and tiny clams (up to half an inch long).

During the stagnation periods between turnovers, life frequently exists where there is no detectable oxygen. These organisms must be essentially dormant or respire anaerobically. In lakes that are constantly stagnant in the profundal zone because of incomplete turnover, animal populations may be extremely sparse.

The periphyton consists of organisms attached to plants, rocks, wood, and other objects. Chief among them are diatoms, green algae, and blue–green algae, which often cover the substrate with a velvety coat. Other attached organisms include mosses, protozoans, and hydras. In the tangle of periphyton and its associated slimy accumulation, planktonic and pedonic species intermingle so that many of them cannot be strictly characterized as one or the other.

PONDS

A clear-cut distinction between lakes and ponds is particularly difficut to make. Many people think of a pond (called tank in some arid regions) as a small artificial body of water on a farm (Fig. 7–8), an abandoned excavation filled with water, or a small natural body of water. Small size seems to be the most accepted criterion, regardless of other limnological characteristics. But what is small? A workable

FIGURE 7–8 *A farm pond in the Midwest.*

definition of a pond is any body of water from a fraction of an acre to a few acres in size. Generally speaking, it is shallow enough for rooted plants to grow over most of the bottom. Exceptions are a few "bottomless" ponds in sinkholes (ponors) and old quarry pits.

Artificial ponds are widely used by farmers and ranchers to meet specific needs like watering livestock. Ponds may be a useful part of a conservation program to check water runoff and prevent erosion. In rural areas where the population is sparse and certain facilities and services common to urban areas are unavailable, ponds are sometimes maintained for fire protection and often for recreational activities like swimming, boating, and fishing.

A fish pond will not be productive if it is simply an impoundment of water. It is an ecosystem that depends on intelligent management by someone who understands the ecology of both the fishes and the other organisms on which they depend or with which they compete. Good management begins with selecting a suitable location and planning the basin and watershed. The watershed must be stabilized with vegetation to prevent erosion and subsequent filling of the basin. Moreover, the watershed must be large enough to fill the basin in a year or less but not large enough to fill it in less than a month. This is necessary because otherwise fertilizers would wash away without being used. Once the lake is completed, good management involves achieving and maintaining balance by stocking, controlling weeds, fertilizing, and harvesting.

Warm-water ponds are so designated if the surface temperature often reaches or exceeds 85°F during the summer. They provide ideal conditions for supporting a balanced population of bluegills and bass. The bass feed on their own young and the smaller bluegills, and the bluegills feed on small invertebrates like the insects (adult and immature); the invertebrates in turn feed on the algae. Fertilizing to increase the algae produces an increase throughout the food chain. To achieve a suitable environment the water should have a minimum depth of four feet, although deeper water may be required, depending on evaporation rates and temperature. In the South four feet is ample if the water level can be maintained. In the North, where winter freezing occurs, the depth should exceed ten feet to ensure that some of the deeper water will remain unfrozen for the fishes. Cold-water ponds, 70°F or less, are best stocked with rainbow or brook trout.

SWAMPS

Swamps have been defined as places where the soil is waterlogged or covered with shallow water and in which some trees grow. Naturally, the flora and fauna differ according to the prevailing climate.

The Okefenokee Swamp in southeastern Georgia and the northeastern tip of Florida is typical. Because of its large size, it is a good example of a reasonably self-contained ecosystem—that is, most of the interactions are internal instead of external (with the bordering area).

Okefenokee occupies 412,000 acres, most of which is covered with shallow water. The area, like any area of comparable size, is not uniform throughout. Within it are islands of high ground, open lakes, watercourses, and treeless marshland ("prairie"). Eighty percent or more of the area is a swamp.

The swamp (Fig. 1–5c) is forested with cypress, tupelo (black gum), bay, and red maple. Beneath the trees are ericaceous shrubs, brambles (*Smilax*), titi (*Cyrilla*), and cassena (*Ilex*). The islands of high ground are covered by pines, although there are patches of hardwoods. The open marshlands, called prairies, are dotted with little ponds and lakes. They are more like quagmires, yielding and unsettled, mostly underlaid with peat. Predominating plants are grasses (including maidencane, of the genus *Panicum*), sedges, ferns, pipeworts (*Ericaulon*), golden-club (*Orontium aquaticum*), and waterlilies (especially spatterdock).

The magnitude of Okefenokee and its various habitats make it a magnificent wildlife refuge. Biologists have studied it extensively and listed many groups of animals. These include 34 fishes, 11 salamanders, 21 frogs and toads, 10 lizards, 34 snakes, 13 turtles, 2 crocodilians (one introduced), 212 birds, and 41 mammals. Twenty additional birds are listed as rare or accidental in occurrence.

The Okefenokee National Wildlife Refuge encompasses about 80 percent of the swamp. Its primary objective is to keep Okefenokee undisturbed, preventing its exploitation for lumber and preserving its sheer beauty for man's enjoyment and to provide a home for wildlife.

Formerly considered wasteland, desolate swamps and marshes are often valuable as nurseries for wildlife or reservoirs of water that recharge aquifers. The recognition of these functions in the ecosystem will stimulate resistance to land developments that would exploit the last frontiers.

SALT LAKES

Many substances in the earth's crust are dissolved by water and add to its impurity. Many of these substances are salts. The oceans have accumulated salts over the millennia in small quantities transported by the freshwater streams that feed them; however, some inland lakes in arid regions are much saltier because of a low volume of water, a high rate of evaporation, and the composition of the rock strata traversed by feeder streams. Even some of the streams in arid regions are too saline for human or agricultural use (Fig. 7–9).

The drainage systems in arid lands transport only a small volume of the water available for runoff, which is further reduced by high evaporation rates. What little water remains accumulates in low places, where it is not of sufficient volume to spill out of its basin and cut a channel to the sea. Instead, it stands and evaporates, leaving its solutes to accumulate. When input and evaporation are approximately equal through the years, the result is a salt lake or perhaps a salina (salt marsh). When evaporation exceeds input the standing water evaporates, leaving the floor of the basin covered with deposits of "salts." The encrusted basins are called playas, alkali flats, or salt flats. Since rainfall fluctuates, a salt lake varies in salinity in proportion to its volume of water, and playas and flats may temporarily become shallow pools or lakes.

The saturation point of water is reached when its salinity is slightly above 25 percent. The salinity of the Great Salt Lake (Fig. 7–10) varies from about 14 percent during times of high rainfall to

FIGURE 7–9 *Bed of the Cimarron River in northern Oklahoma. In this segment of the river the water is very salty. Unfit for agricultural plants, the soil supports salt cedar* (Tamarix), *a salt-tolerant plant.*

FIGURE 7–10 *Shoreline of Antelope Island in the Great Salt Lake. The beach, a mixture of sand and salt, is covered with dead brine flies. (Photo by C. R. Wylie.)*

27 percent during droughts. The average is about 25 percent. The Dead Sea, which is really a lake, has a salt content averaging about 25 percent but can become chemically saturated in places. Lake Van in Turkey has an even higher concentration of solutes and is reputed to be the most saline lake in the world. In fact, many smaller bodies of water evaporate to the point of saturation and beyond to the point of precipitation.

The words salt and alkali are loosely used to describe the mixtures of chemicals that accumulate in inland basins. Salts present around in the Dead Sea are sodium chloride, magnesium chloride, calcium chloride, potassium chloride, magnesium bromide, and calcium sulfate. Quite often water in salt lakes like the Dead Sea and the Great Salt Lake contains more sodium and chloride than any other dissolved substances and is definitely salty tasting. By contrast, Lake Van contains carbonates of soda, borate of sodium (borax), chlorides, and sulfates that make the water bitter to the taste—that is, alkaline rather than salty. Whatever the chemical concentrates, environments like these are inhospitable to all but a very few specialized species.

Life in salt lakes is confined to a very few species. The Dead Sea is completely devoid of animal life, and the Great Salt Lake has no fishes. Contrary to expectations, Lake Van abounds with a species of herring. The Great Salt Lake, though fishless, has a few other kinds of life, including a blue–green alga, bacteria, two flagellates, two brine flies, a brine shrimp, and a few protozoans. In the marshes around the lake and on the islands are waterfowl—California gulls, Caspian terns, white pelicans, Treganza great blue herons, and double-crested cormorants. In small pools in Death Valley, where the salinity is at least six times that of the ocean and the water temperature may reach 112°F, there are small pupfish. Moreover, blue–green algae grow in some lakes in California and other places where the water is so briny that the salt crystallizes.

REFERENCES AND FURTHER READINGS

Bennett, G. W. 1962. *Management of Artificial Lakes and Ponds*. Reinhold, New York.

Frey, D. G. 1963. *Limnology in North America*. University of Wisconsin Press, Madison.

Hynes, H. B. N. 1960. *The Biology of Polluted Water*. Liverpool University Press, Liverpool, England.

International Symposium on Eutrophication. 1969. *Eutrophication: Causes, Consequences, Correctives*. National Academy of Sciences, Washington, D.C.

Langlois, T. H. 1954. *The Western End of Lake Erie and Its Ecology*. Edwards, Ann Arbor, Mich.

U.S. Department of Agriculture. 1955. *Yearbook of Agriculture: Water*. GPO, Washington, D.C.

Welch, P. S. 1952. *Limnology*. 2d ed. McGraw-Hill, New York.

Chapter 8
Oceanic and Tidal Environments

MAGNITUDE OF OCEANS

The oceans comprise nearly all of the hydrosphere. Their outstanding features include vastness, basin configuration, stabilizing influences, absorbing capacity, and internal mixing. Although about 70 percent of the earth's surface is marine, the volume of the oceans is even more startling. The ratio of the volume of water held in the oceanic basin to the volume of land projecting above sea level is about 95 percent water to 5 percent land. Basin configuration includes shoreline and bottom topography. The shoreline (intertidal zone as used here) and the shallow water adjacent to it are richest in life forms because of the inflow of nutrients from the land and the availability of light for photosynthesis through much of its depth.

Throughout geological history the shoreline has shifted when either the land or the water has moved up or down with respect to the other. The crust of the earth is by no means as durable and unyielding as it seems. Under pressure, the hard earth substances are plastic enough to bend or shift position. Even if the continental masses never changed, the water level could fluctuate because major changes in the level of the ocean are in direct proportion to the amount of water trapped in polar ice caps and glaciers. When glaciation increases, the level of the ocean drops; when the ice melts, the level rises. Any shoreline sinking with respect to the water level is drowned and irregular; if it is rising, it will be planed by wave action and become regular. The drowned shoreline provides a greater variety of habitats.

The ocean floor has larger irregularities than the continental landmasses. Besides mountain ranges, there are volcanic cones and guyots, extensive plains, plateaus, continental shelves, rifts, and trenches. Some mountains that project from the ocean floor are higher than any of the peaks on the continents but are so deeply submerged that knowledge of their existence had to await the development of modern exploratory techniques. Even now, that vast and barely accessible domain remains poorly understood.

CHEMICAL COMPONENTS

The salinity of most oceanic waters is approximately 3.5 percent. Where evaporation is higher, salinity increases—in the Red Sea it is 4 percent. Similarly, freezing increases the salinity of oceanic water because only the solvent is removed from the solution, leaving saltier water behind. The latter, being denser, sinks. Actually, much of the polar ice comes from precipitation, but there is enough freezing of polar oceanic waters to increase salinity slightly.

A countercondition leading to decreased salinity exists at the poles. Because water expands by 9 percent when it freezes, the resulting ice floats at the surface, where it receives maximum warmth from the sun. As the ice melts, fresh water is added back to the oceanic reservoir from the continents.

In no place does the salt concentration of the oceans approach the salinity of some landlocked bodies of water such as the Great Salt Lake, which is sometimes as much as 27 percent salt on a weight–volume basis. Salinity is reduced where there is local dilution during rainstorms or along coastlines where continental drainage flows into the ocean. The dilution is small except around the mouths of rivers. Ocean water is, in general, kept uniform by the constant mixing caused by currents, waves, tides, sinking, and upwelling.

Along with salt, many elements and compounds are transported to the oceanic reservoir from the continents either by water or by air. Some of the materials are dissolved and others suspended. Many are valuable but are in such low concentrations that they cannot be economically separated from the water. A few that are profitably extracted under some circumstances are sodium chloride, bromine, and magnesium. Now that water itself is needed so badly in many places, important research efforts are under way to discover more economic ways to extract fresh water from sea water.

Living organisms are capable of accumulating substances through metabolic processes. Many marine animals take in calcium carbonate and use it for bones, shell, or other supporting and protective structures, and some algae precipitate it as an encrustation around their bodies. Some organisms like diatoms, radiolaria, and one group of sponges secrete supporting walls or skeletons of silica. Even gold is known to be concentrated by certain seaweeds from the minute quantities dispersed in ocean water, about 0.0004 ounce per million gallons. Among the seaweeds, the browns, rich in potassium and iodine, are sometimes used for fertilizer by Europeans and Asiatics. Blue–green algae, which could be collected during blooms, have been suggested as a source of fertilizer because of their high phosphorus and nitrogen content. Finally, seafoods are well-known dietary sources of iodine.

Conditions do exist in the oceanic environment that allow the deposition of specific substances. For example, large deposits of limestone, dolomite, and phosphate rock are produced by chemical precipitation, biological concentration (e.g., shells), and evaporation of water. Most deposits probably result from small changes in the solubility of the minerals in water. Since the water is nearly saturated with these minerals, a small change in pH may result in supersaturation and the formation of crystals, which then settle to the bottom. The pH of water is largely regulated by the carbonate, bicarbonate, and dissolved carbon dioxide content. In bodies of water where there are large quantities of phytoplankton, photosynthetic activity may be enough to reduce the carbon dioxide content and effect a sufficient change in the pH to cause mineral precipitation.

Not so clear are the reasons for the mineral nodules abundantly scattered over some parts of the ocean floor. These nodules of iron, nickel, manganese, and cobalt can be several inches in diameter and may represent an important natural resource. Several kinds of **oozes** also collect on the floor. They consist of atmospheric dust and organismic remains. One such ooze, derived from diatom walls, has been uplifted and converted into dry-land deposits that are mined and used for abrasion, filtration, and insulation.

Less obvious but perhaps much more important than the solutes and dispersed particles found in oceanic water are the dissolved gases. Of particular interest are the oxygen and carbon dioxide derived from the metabolic processes of respiration and photosynthesis. On a short-term basis they seem to remain at a steady level relative to each other. Obviously, the ocean can interact with the atmosphere by simple diffusion. In this respect it acts as a reservoir that either absorbs or loses gases through the air–water interface, depending on the concentration of each substance in the air and water. Although this mechanism has a stabilizing effect, wide swings in atmospheric composition have occurred in the past, as evidenced by intermittent bursts of plant growth clearly preserved in the fossil record.

TEMPERATURE

Oceanic temperatures are much less extreme than air temperatures. Surface temperatures vary from about 28°F at the poles to 86°F in the Persian Gulf. All deep bodies of water are colder at the bottom than at the surface except when a temperature inversion occurs (Chapter 7). In oceans there is a temperature decline beginning at 100 to 600 feet, and it becomes uniformly cold, near freezing, at about 3000 feet below the surface. This transition zone in which the temperature drops rather quickly is comparable to the thermocline of deep lakes.

CURRENTS

Salinity and temperature, as well as wind, earthquakes, rotation of the earth, and gravitational effects of the moon and sun, keep the water in motion—in both continuous patterns and unpredictable turbulence. The latter often produces havoc when it cannot be contained and sweeps ashore. A major tragedy of this kind occurred in November, 1970, when a cyclone and tidal wave bowled into East Pakistan. Estimates put the immediate loss of human life at a quarter of a million and subsequent loss from starvation at an even higher figure. Tides and waves likewise pound the land, making their greatest impact in shallow water in general and the intertidal zone in particular. Their cause and effects will be discussed later.

The rotational currents of oceans consists of surface streams and convections. Naturally, the surface currents are better known (Fig. 8–1). Most obvious are the gigantic oceanic whirlpools whose outside margins turn faster than their centers. The margins or rims flow like big rivers. A rim is made up of segments called currents or streams, each with its own name. The best known of these is the Gulf Stream, which comes from the Gulf of Mexico and flows north as it skirts the eastern seaboard before heading eastward across the Atlantic just south of Newfoundland. About halfway across the ocean, it splits into two currents, identified by different names. Around Florida, where the Gulf Stream is about 50 miles wide and 1500 feet deep, the current flows at about 5 miles per hour.

A simplified explanation of the whirlpool action of surface currents is that they are the combined effect of temperatures, winds, and the spin of the earth. At the equator, water heats and expands, making it a little higher than the water near the poles. Equatorial water flows to the polar region to replace the heavier, colder water that sinks to the bottom. The spin of the earth makes water pile up several inches higher on the oceans' western rims than on their eastern rims. Superimposed on all the surface currents is the **Coriolis effect**, which also results from the spin of the earth. It deflects the

FIGURE 8–1 *Major currents of the oceans of the world. Note that the currents rotate clockwise in the Northern Hemisphere and counterclockwise in the Southern Hemisphere. (Based upon data of the U.S. Navy Oceanographic Office. Reprinted with permission of the author and publisher from Arthur N. Strahler, 1960,* Physical Geography, *John Wiley & Sons, New York. ©1960, John Wiley & Sons, New York.)*

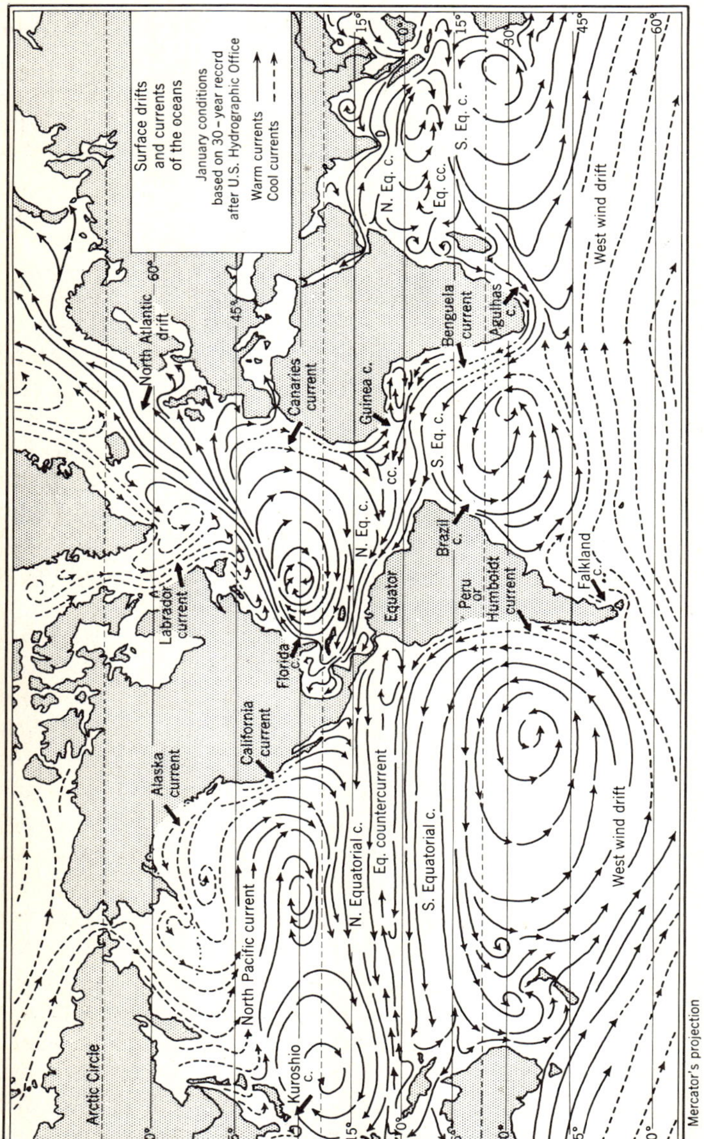

currents toward the west if they are near the equator and toward the east if they are near the poles. Finally, air currents that result from the same Coriolis effect push the water currents along as they move against the surface. Penned between landmasses, the deflected currents merge into the familiar circular pattern, clockwise in the Northern Hemisphere and counterclockwise in the Southern.

Some of the surface water flowing toward the poles is diverted into convectional circulation. What happens is that polar water, heavier because it is colder (down to 39°F) and frequently saltier, sinks to the bottom and is replaced by lighter surface water moving in from nearer the equator. Some of the colder, saltier water moves along the bottom toward the equator, where it apparently rises to replace the surface water flowing toward the poles. How long it takes the water to move from the poles to the equator is still largely a matter of conjecture. Estimates vary from about 300 to 1500 years. Besides moving in the equator-to-pole conventional flow just described, water also circulates in a more local vertical pattern known as **upwelling**.

UPWELLING

The phenomenon of upwelling is important in bringing mineral nutrients to the surface, where they sustain oceanic gardens. This can happen when water rises to replace outflowing surface water, heavier water sinks to replace lighter water (at the poles), and currents of lighter water encounter heavier water. The turnover is slow, although a vertical displacement of sixty-five feet per month has been recorded off the coast of California. The enriched water from below favors both plants and animals concentrated close to the surface. In particular, it stimulates diatoms and dinoflagellates, abundant photosynthesizers of the plankton. The tiny diatoms, which thrive in colder waters, develop blooms of growth that literally color the ocean yellow, brown, or green. At such times there may be tons of the minute plants per acre. The bonanza of food attracts animals by the billions, and these locations become excellent fishing grounds.

Large upwellings occur along coastlines, especially on the western margins of continents. The largest are along the middle third of the coastlines of both North and South America and along the outer thirds of the west coast of Africa (Fig. 8–2). Others are around the border of Antarctica, fringing the northwestern tip of Australia, outside the Chesapeake and Delaware Bay area, off Muscat and Omanin in the Arabian Sea, and off the Somali Republic at the eastern tip of Africa.

FIGURE 8–2 *Locations of major oceanic upwellings along the margins of landmasses.*

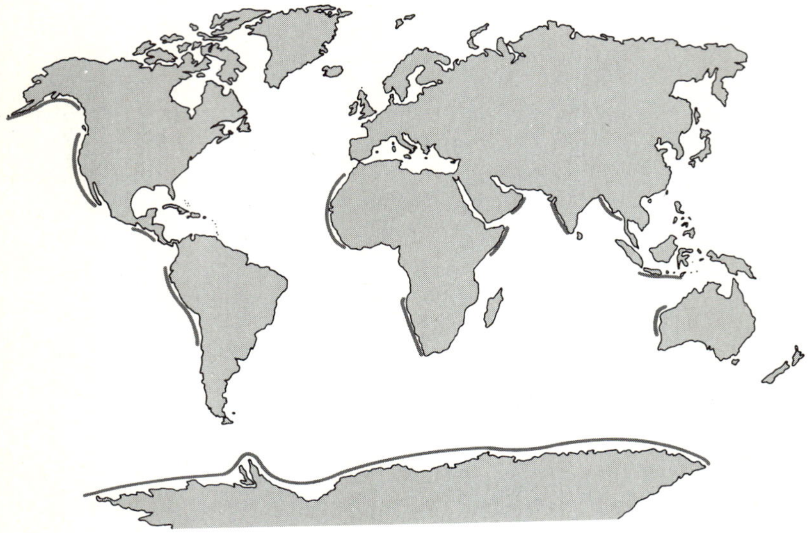

MARINE ENVIRONMENTS

Although the oceans are much larger than the continents, their biotas do not lend themselves as well to compartmentalization into biomes. This is due in part to the nature of water itself: Its temperature is much more constant than that of the air. Because of its uniformity, boundaries between habitats are generally indistinct, and organisms are correspondingly less variable. Yet, unspecialized as oceanic life in general seems to be, it is sometimes surprisingly intolerant of small environmental variations, which produce catastrophic "die-offs." The plants that are the primary shapers of terrestrial biomes have no oceanic counterparts. This is not to imply that plants are not important, for in the oceans too they are always the primary producers of food. Most, however, are small, simple algae (diatoms, dinoflagellates, blue–greens, and seaweeds), hardly influencing animals except by serving as food and having little or no effect on the surroundings. The oceans are the domain of animals, not plants, and many of them seem to move around with greater freedom than do land animals. There is no doubt that many of the fishes are much less restricted in their ranges than most land animals. Finally, the classification of marine communities is handicapped by inadequate knowledge due to the vast size of the oceans and the difficult technical problems involved in its exploration.

To facilitate the discussion of marine habitats, certain terms that have gained wide usage need to be understood. Specific types of environments (Fig. 8–3) are designated **littoral** (between high and low tides), **pelagic** (ocean water), and **benthic** (ocean floor). The same adjectives may be used to describe the animals or plants living in those environments—for example, littoral crabs, pelagic mammals, or benthic gastropods. These locations are sometimes subdivided in order to be more specific; the pelagic environment, for example, is appropriately divided into photic and aphotic strata, depending on whether or not there is sufficient light for photosynthesis.

What is the best way to study marine organisms and environments? Should the entire ocean be treated as a single large biome,

FIGURE 8–3 *Major oceanic environments (depths not proportional).*

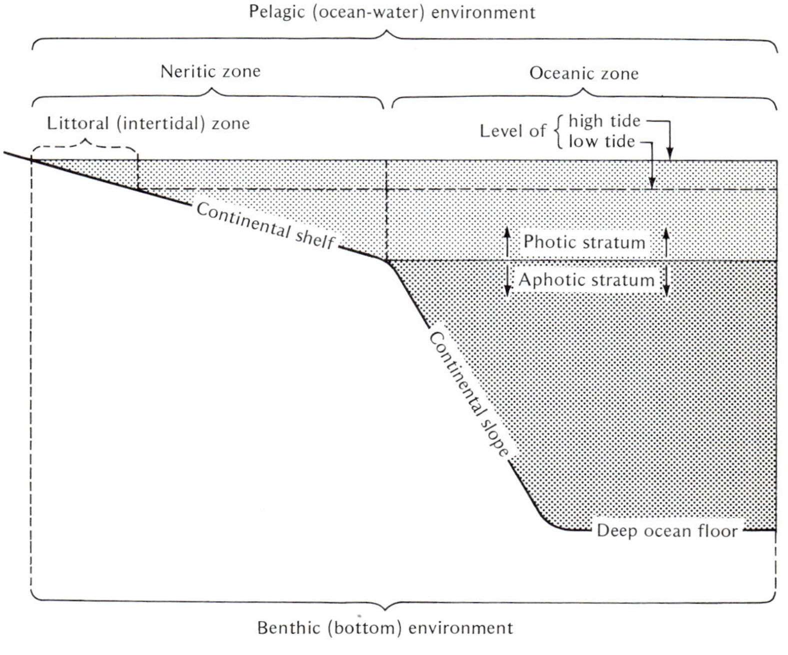

as several distinct biomes, or in some other way? Water and temperature, the two variables mainly responsible for the different terrestrial biomes, are comparatively uniform in the ocean. Water deficiency can exist only under very limited conditions where salinity is abnormally high (physiological drought). Daily, seasonal, and latitudinal temperature ranges are moderate. In view of fundamental differences in terrestrial and marine environments that make it difficult to apply the biome concept, the most appropriate solution is to look at large zones or divisions that have some unifying characteristics. On this basis, the ocean is divided into the following locations occupied by distinctive flora and fauna:

- Intertidal (littoral) zone
  - Sandy beaches
  - Rocky beaches
  - Salt marshes
  - Tidal pools
- Shallow ocean
  - Continental shelf
  - Banks
  - Reefs
- Deep ocean

INTERTIDAL (LITTORAL) ZONE

TIDES

The **intertidal zone** is the coastal strip alternately flooded at high tide and exposed at low tide. At this meeting of land and sea, the boundaries are set by the tides, but within these boundaries wave action is an important additional influence. Together the tides and the breaking waves make this habitat one of instability and stress.

The tides are bulges in the oceans caused by the gravitational pull of the moon and the sun. The proximity of the moon gives it over twice the gravitational pull of the sun. At new and full moon, when the moon and the sun are in line with the earth, they act together to cause the highest tides, called **spring tides**. When the moon and the sun are at right angles to each other, as at the first and third quarters of the moon, they pull from different directions so that their effects are counteracting, the net effect being the lowest tides, called **neap tides**. Tidal bulges occur simultaneously on opposite sides of the earth. At new moon, when both sun and moon are pulling together on the same side of the earth, the bulge on the opposite side is left behind, so to speak, as the earth is pulled away from it. To understand this phenomenon, one should realize that the solid geosphere, as well as the hydrosphere, responds to gravitational effects and that the force of the attraction between the earth and the heavenly bodies is related to the distance between them.

Based on the movement of the moon, the tidal bulge should return every twelve hours and twenty-six minutes, but this is not necessarily the case. For instance, in some parts of the Pacific and Indian oceans the tides come only once a day. The rhythm and magnitude of tides are strongly modified by the uneven distribution of the tidal force, caused by local topography and the size of the basin. Each locality is unique. To illustrate, take the case of a steep-sided, long inlet with a wide exit to the sea—in essence, V-shaped. A tidal bulge entering the opening of the V is confined to a narrowing channel and increases in amplitude as it moves along. The Bay of Fundy, famous for this phenomenon, has tides sometimes over forty-five feet high. In contrast are the weak tides of less than a foot in the Gulf of Mexico, whose basin has a different shape.

WAVES

Life in the intertidal zone, and on shallow continental shelves as well, receives a pounding by waves unlike the gentle wash of tides. Waves are driven by winds and are proportional in size to wind velocity and unobstructed distance. They are surface features, seldom perceptible below twenty to thirty feet except in stormy seas, when the turbulence may reach a depth of 500 feet or more, even to the bottom of the continental shelf. Wave motion sharply diminishes below the surface turbulence. The water of a wave does not move forward; only the wave form moves. The water merely rotates in a circular pattern, coming to rest almost where it started. It is the breaking of the waves along the coastline that subjects the animals and plants to serious danger as the water erodes here and deposits there, continually changing the face of the bottom. Organisms respond accordingly; they attach firmly or hide beneath the surface to avoid being battered.

The breaking of waves is caused by their frictional drag on the bottom. The point at which a wave drags the bottom is related not to its height but to its length, the distance between the crest of one wave and the crest of another. Waves generally drag the bottom when the depth of the water is about half the length of the waves. Normally, they can reach a height of about one-seventh the distance between crests before collapsing as whitecaps.

PERIODICITY OF THE INTERTIDAL ZONE

The intertidal zone is definitely a habitat of extremes. Within the short time span of six hours and thirteen minutes, the zone changes from total inundation by water to total exposure to air. As the tide ebbs, floaters and swimmers follow it out, and most bottom dwellers that can move about hide under something or burrow in the ground where they can protect themselves from sun and wind. There they keep their respiratory organs moist and await the return of the tide. Land animals seize the opportunity to forage over the exposed bottom (beach) for hapless animals left stranded in defenseless positions or to probe for living and nonliving tidbits embedded in the sand. In particular, birds, spiders, insects (e.g., beetles and ants), and crabs make the most of the periodic restocking of their hunting ground with food brought by the tides.

The kind of life most abundant at any particular place in the intertidal zone depends to a great extent on the nature of the bottom—sand, mud, or rock. Ordinarily, people envision the zone as it appears when the tide is out; they consider the exposed bottom and its inhabitants and ignore the oceanic portion that periodically covers it. The floating and swimming organisms of the intertidal zone are discussed in the next chapter.

The difference in living conditions between the high-water mark and the low-water mark is great in several respects. Near the high-water mark the water is out most of the time, leaving the bottom exposed to drying and extremes of temperature; near the low-water mark the area is flooded most of the time, making conditions more uniform. Sometimes these differences in length of exposure, plus other factors like slope or composition of the bottom, result in distinct floral and faunal zonation. The intertidal zone may be clearly divided into upper, middle, and lower—often called upper, middle, and lower littoral zones. The land immediately bordering the upper zone is the supralittoral zone, whereas the bottom adjacent to and below the lower zone is the sublittoral zone (continental shelf).

SALT MARSHES

Salt marshes (Fig. 1–5a), mud flats, and mangrove forests (Fig. 8–4) are usually broad, low-lying zones around estuaries where alluvium settles out of the fresh water arriving from inland sources. The bottom is shaped by the streams, tidal wash, and plants. Since there is a mixture of fresh and salt water, salinity varies considerably from place to place.

FIGURE 8–4 *A mangrove forest in southern Florida. Note the prop roots under the branches of the trees in the foreground. (Photo by G. C. Smith.)*

Until recently, salt marshes were considered virtually useless from an economic or biological point of view. Now, enterprising land developers are attempting to fill these vast acreages and turn them into homesites where the owner can have a car in the front yard and a boat in the back. Biologists and others are just as energetically resisting this trend. Such areas must be preserved in an unaltered condition as breeding grounds for a number of important marine organisms like shellfish, shrimp, and fishes. They are teeming with annelids, clams, oysters, fiddler crabs, young fishes, and birds.

The extensive salt marshes of the South Atlantic coast look somewhat like savanna or prairie because there are dense stands of salt grass (*Spartina spp.*) and rushes (*Juncus roemerianus*). These grow thick and tall (four or five feet). Other zones of vegetation occur at slightly higher elevations. These include other grasses, sedges, succulents (*Salicornia*), and showy flowers like marsh morning glory (*Ipomoea sagittata*), marsh aster (*Aster tenuifolius*), and seashore mallow (*Kosteletskya virginica*).

SANDY BEACHES

Life on sandy beaches is visibly dominated by animal forms, although microscopic algae and bacteria are found in the sand, and seaweeds may sometimes occur in tidal pools. Some of the micro-

FIGURE 8–5 *Drift line on a sandy beach where animals tend to congregate in search of food.*

scopic animals living in the sand are copepods, ostracods, nematodes, and gastrotrichs. The tiny beach fleas (sand hoppers) and the large ghost and mole crabs are burrowing crustaceans. On both coasts is the tiny sand-burrowing clam of the genus *Donax*. Other bivalves like the heart shell (*Cardium*) and razor clam (*Tagelus*) may be buried in the lower beach. There may even be some sandworms beneath the surface. The nocturnal ghost crabs scavenge along the drift line (Fig. 8–5) and dart sideways into their burrows when danger approaches. As a habitat the sandy beach is rich in neither variety nor number.

The outstanding feature of a sandy beach is its instability. The constant agitation by water, the alternate flooding and drying, and the shifting of sand particles subject beach life to extreme hardships. The problem of protection is often solved by hiding below the surface, as illustrated by the animals just mentioned. Since the beach environment is unfavorable for plants, there is also a paucity of food for animals.

ROCKY BEACHES

Rocky beaches (Fig. 8–6), jetties, and breakwaters are much more resistant to wave action and currents than the unconsolidated particles of sandy beaches. There, the firm substrate is unsuitable for the burrowing animals typical of sandy beaches but ideal for attached plants and animals that can withstand pounding by the sea.

Many algae, especially brown types like the kelps and rockweed (*Fucus*), grow in large beds attached by a specialized base known as a holdfast. They are tough plants; otherwise they could not survive direct exposure to the sunlight, wind, and extremes of temperature that prevail when the tide is out. Lichens and blue–green algae, also tough plants, are often found in the upper zone and adjacent supralittoral border.

The most obvious animals are acorn barnacles, clams, and oysters—firmly attached and growing close together. Each of these animals lives in its own little "microhabitat," protected in its calcareous shells. Barnacles and oysters are cemented to rocks by their shells, but clams are attached by byssal threads, tough horny fibers produced by the foot of the body. Different types of sponges may also attach themselves to the rocks. Motile animals live among the sessile ones or hide among seaweeds. Common animals living on rocky beaches are annelids, springtails, starfishes, sea urchins, and crabs.

FIGURE 8–6 *Rocky beach in Maine. Organisms in this habitat must be firmly attached to withstand the pounding of the waves. (Photo by L. P. Stratton.)*

TIDAL POOLS

Tidal pools (Fig. 8–7) are bodies of standing water left behind when the tide is out. Most are small and shallow, but some are large and deep. More extreme environmental conditions develop in the smaller ones, where the water becomes much warmer or colder than in the ocean. Normally, evaporation is accelerated and salinity increased, sometimes to such an extent that salt crystals precipitate. The pools may, however, be less saline during times of heavy rainfall. If algae abound, their photosynthetic activity modifies the normal oxygen–carbon dioxide ratio. The wide fluctuations of the environmental factors within tidal pools make some of them intolerable except for the hardiest species.

Most occupants of tidal pools are probably stranded there by the ebbing tide, although a few organisms select specific pools for one reason or another. The more permanent pools may have well-established populations of harmonizing species. Whether the basin is rocky or sandy has a great deal to do with what kind of community is present. Several kinds of algae do very well in some pools. Animal life in tidal pools often includes protozoans, sponges, polyps, worms, chitons, mussels, snails, starfishes, brittle stars, crabs, barnacles, springtails, tunicates, and many other species.

FIGURE 8–7 *A tidal pool in Hawaii. In the pool are sea urchins, a sea cucumber, and algae. (Photo by G. W. Fairbanks.)*

FIGURE 8–8 Salicornia, *a succulent seed plant, and the openings to fiddler crab burrows in a coastal area.* Salicornia *also grows in saline sites in arid regions.*

Tidal pools may dry up completely, leaving a salty encrustation on the bottom. This is likely to occur if they are located nearer the high-water mark, where drying occurs for longer periods. In salt marshes these **salt pans** may have in or around them the succulent glasswort (*Salicornia*) (Fig. 8–8), which often catches the eye with its red coloration as it becomes older. Other plants may grow in zones along the periphery.

REFERENCES AND FURTHER READINGS

Engel, L. 1961. *The Sea*. Time-Life, New York.
Kucera, C. L. 1973, *The Challenge of Ecology*. Mosby, St. Louis, Mo.
Marx, W. 1967. *The Frail Ocean*. Ballantine, New York.
Nybakken, J. W. 1971. *Readings in Marine Ecology*. Harper & Row, New York.
Odum, E. P. 1971. *Fundamentals of Ecology*. 3d ed. Saunders, Philadelphia.
Thorson, G. 1971. *Life in the Sea*. McGraw-Hill, New York.

Chapter 9
Oceans

SHALLOW OCEAN

CONTINENTAL SHELF

Sometimes the edge of a continent continues out to sea as a gentle slope or shelf for many miles before abruptly plunging two or more miles. Most shelves are the shallow portions of continental slopes upon which sediment from inland sources is deposited; the sediment is rearranged by wave action. The water at the outer edge of the shelf is about 1500 feet or less in depth, while much of it is about 400 feet deep. The shelf along the eastern coast of North America is often 100 or more miles wide but is not as wide as some shelves around the Arctic Ocean or along the northern coast of Australia, which are from 800 to 1100 miles wide. Extremely narrow shelves are found along the western borders of the Americas and Africa.

That part of the ocean of which the shelf is the floor is the **neritic zone**. It is occupied by both pelagic and benthic organisms. Since the water is relatively shallow and enough light to support photosynthesis can penetrate to the depth of 300 to 400 feet, algae can grow all the way to the bottom in some places. Most photosynthetic activity, however, is limited to depths of 200 feet or less. Plant growth is also favored by nutrients washed in from continental drainage. At the same time, both plants and animals benefit from the moderate temperatures of the shallow water above the shelf. These favorable conditions make the neritic zone one of the most densely populated oceanic habitats. Similar advantageous conditions prevail where there are banks or coral reefs.

Oceanic plants of the continental shelf are primarily algae—the seaweeds and an assemblage of smaller diverse forms. Giant kelps on the Pacific Coast are attached as deep as 360 feet, and some red algae are found in even deeper waters. The seaweeds are normally attached but often break away and float, sometimes in large quantities. Perhaps the best known is the gulfweed *Sargassum*, which has berrylike, gas-filled bladders that give it buoyancy. The detached plants are swept along by the surface currents that encircle the North Atlantic like a gigantic whirlpool. They accumulate in its "eye" in an area of about two million square miles (approximately 2000 miles long by 1000 wide). This unique location has been named the Sargasso Sea after the conspicuous seaweed. Many small algae, together with small larvae and adult animals, form the neritic plankton, the important food of larger forms of life. The composition and distribution of plankton are discussed in the section on deep ocean.

While algae are the predominant marine plants, a few sea plants may grow in the shallow oceanic water. One is eelgrass (*Zostera*), which grows in thick beds in colder estuarine waters along the seacoasts of both the Northern and Southern Hemispheres. Its leaves are ribbonlike, many times longer than wide. The North American variety has leaves up to 0.2 inch wide and 1.5 to 3 feet long. The near-destruction of eelgrass by a parasitic slime mold (*Labrinthula*) on the Eastern Seaboard in 1931 focused attention on how important a single species can be in the economy of a community. Either because the eelgrass was a direct link in the food chain or because it provided a necessary habitat, some animal populations declined sharply. The sea horse almost disappeared. Other species seriously affected were scallops, crabs, flounder, cod, and even brants (geese). In time, the eelgrass recovered to a considerable extent and along with it the associated animal populations.

Almost all major groups of animals are found on or above the shelf. The invertebrates dominate the floor and fishes abound in the water above it. Wherever a suitable substrate exists, sponges, hydroids, sea anemones, corals, bryozoans, pelecypods, barnacles, or sea squirts (tunicates) may be found in remarkable abundance. Moving around on the bottom are scallops, sea snails, crabs, horseshoe crabs, starfishes, sand dollars, sea urchins, sea cucumbers, rays, toadfishes, and flounders. Large numbers of annelids and molluscs burrow in the bottom and extract food and oxygen from the water which they circulate through their siphons or tubes. Some annelids expose only their respiratory "gills" and mouth beyond the safety of their burrows. Commercial fishes of Northern Atlantic waters include cod, halibut, herring, haddock, and mackerel. Farther south fishermen take menhaden, mullet, bluefish, bass, and others. Swimming animals besides the fishes are cephalopods, shrimps, and several kinds of mammals like seals and porpoises. Larger floating animals are jellyfishes and Portuguese men-of-war.

BANKS

On the continental shelves there are shoals or shallows known as **banks**. Some of them are in water only twenty to thirty feet deep, but others are in considerably deeper water. They are among the best fishing grounds in the oceans because they are favorable sites for autotrophic plants, the direct or indirect source of food for all animals. Near the beginning of the food chain are tons of **zooplankton** that, together with the phytoplankton on which they live, provide a food base for the large populations of fishes congregating there. There is good reason to believe that the zooplankton moves up and down through most of the water over the continental shelves.

The Grand Banks southeast of Newfoundland are among the best such areas. These shallow banks are 500 miles long and submerged 50 to 1000 feet under water. The cold Labrador Current (33°F) descends from the north and the warm Gulf Stream (over 50°F) sweeps up from the south, each bringing a different set of organisms attuned to its unique conditions. More than anything else, this convergence of two sources of planktonic food in shallow water accounts for the abundance of fish food there. Among the fishes taken on and around the banks, as in other northern waters, are cod, halibut, herring, haddock, and mackerel.

REEFS

Reefs are stretches of stony corals and associated organisms or islands built on their remains. Corals grow in warmer seas where the minimum temperature is seldom below 68°F and the salinity is only slightly diluted by fresh water from landmasses. They grow in shallow water, rarely below 150 feet, but must be submerged at low tide or constantly splashed by waves. They grow best in clear water where their delicate polyps are not subjected to harmful pollutants and light is not obstructed. Light is important for species of phytoplankton used directly or indirectly for food and for algae that live symbiotically in the tissues of many corals, presumably to the benefit of the corals.

Reefs are spectacular in the color and form of the corals themselves and of the other life forms that are constantly associated with them. The general appearance is something like a flower garden blossoming with delicate tentacle-tipped polyps and tubeworms. Coral polyps are grouped in colonies whose variety of shapes is even more colorful than the individuals of which they are made. They may be shaped like mounds, antlers, mushrooms, rods, spires, wands, and fans—some rigid and others flexible, some simple and others branched, some erect and others sprawling. No other natural community can approach their artistic array.

The matrix of the reef consists of colonies of stony corals that overlap and pile upon others to form a honeycombed and tangled conglomerate. The whole assemblage is more or less cemented together and encrusted with the limy deposits of coralline algae. To this framework are attached colonies of horny corals and individual polyps of sea anemones, both close relatives of the stony corals. The individual polyps of corals live in pits into which they retract for protection. The skeletal material of horny corals is a flexible horny material in the shape of fans or wands (whips). All that is necessary to complete the classic picture of a reef is to put colorful fishes in the foreground and a moray eel or an octopus lurking in a crevice.

Of all the marine habitats, reefs have the greatest population density. Many kinds of animals attach to surfaces, bore into the stony material, occupy natural spaces, crawl over the surface, and swim around or over the branches of coral. The common reef animals are sponges, bryozoans, clams, sea snails, sea urchins, crabs, lobsters, shrimp, starfish, tunicates, and fishes. One of these, the crown-of-thorns starfish (*Acanthaster placi*), is currently bringing about widespread destruction of living corals inhabitating some of the reefs of the Pacific.

Some reef occupants are dangerous to man. One is a giant clam as much as four feet across and weighing 500 pounds. The margins of its two hinged shells (valves) have inverted V-shaped projections like large teeth that can bite into a leg like the jaws of a bear trap. Some reef inhabitants are poisonous—a few sea urchins, sting rays, a few bony fishes, and sea serpents. Poisonous fishes inject their poison with spiny fins. Examples are the scorpion fish, the lion fish, and the stonefish, the latter being the deadliest of all reef creatures. A few other fishes are dangerous because of their bite. The moray eels (some venomous) may inflict damage but probably attack only when endangered. Dangerous predators from the open sea also frequent reefs: sharks, barracudas, and giant groupers.

Reefs occur in the tropics or subtropics, or even in some distant nontropical places that are washed by warm currents like the Gulf Stream. As previously mentioned, they may be alive and growing or ancient remains, now the "bedrock" of uplifted islands. Therefore, the term *reef* applies to any geographical feature of coral origin. The living reefs grow at an estimated rate of about one foot every 300 to 400 years. Because they are so close to the surface, they are ever-present hazards to ships. Submerged or emergent (islands), they are classified as fringing, barrier, or atoll, depending on their relationship to present or ancient shorelines.

Apparently the atoll terminates a sequence of events beginning with the birth of a volcanic island above the oceanic surface. Along its margin or anywhere else where the water is shallow enough, reefs become established. According to the **glacial control hypothesis**, the sea level was lowered during the ice age when a great deal of the earth's water was locked in ice on the continents. During this time coral reefs, and other islands as well, were truncated by wave action. As the ice melted and the level of the sea began to rise, corals grew along the fringes of the islands; their growth kept pace with the rising water. The fringes of corals now surround central lagoons about 300 feet deep, probably indicating the level to which the surface of the ocean was lowered during the ice age.

A reef probably becomes the foundation of dry land through either the uplift of land or the lowering of the sea. At any rate, some plants like the mangroves become established whenever an opportunity exists. Their salt-tolerant embryos, partially projecting from the seed pods, are washed about by waves and currents; some of them eventually lodge in crevices where they can get a start. Once established, nothing is better than their interlocking roots for collecting debris and mineral particles wafted by waves and tides. This is the first phase of soil building, which covers the corals and prepares the way for succession of plants and animals. The corals die, except in marginal waters, and their skeletal remains are never seen except when excavated. Coral islands are often covered or fringed with coconut groves that first got started from fruits floating there on oceanic currents.

Of all reefs, the largest and best known is the Great Barrier Reef off the northeast coast of Australia. It is a rather continuous unit consisting of submerged coral beds, sand bars, and islands among which are serpentine passages. Even within the large barrier system, there are subunits classifiable as fringing reefs, barrier reefs, or atolls. Altogether, the system is over 1200 miles long and covers about 80,000 square miles. The great reef is located a mean distance of about thirty miles from the shore and is separated from it by a lagoon 100 to 400 feet deep.

DEEP OCEAN

PHOTIC STRATUM

The best way to deal with deep oceans is to look at their stratification. Since light cannot penetrate deep enough to support photosynthetic activity in plants on the bottom, they can exist only if they float on or are suspended near the surface. Most of the photosynthetic plants are probably concentrated in the upper 100 feet. By far the most numerous and most important of these are the diatoms, tiny unicellular golden-brown algae. Since they contain chlorophyll, they are the major producers at the beginning of the food chain. They are easily more numerous than all associated organisms combined. The dinoflagellates and other flagellates, also photosynthetic, are likewise important in the oceanic pasture. With them are several less important types of small algae, bacteria, and other fungi. Feeding on these plants are a host of immature and adult animals, most of which are minute. This potpourri of small plants, eggs, and animals is the group known as plankton.

Individuals of the plankton float or swim, but the swimmers, wafted about by waves, are unable to make headway like the large fishes or mammals. In addition, they make a diurnal up-and-down migration. During the day they retreat to deeper water where they are protected from intense sunlight. High-frequency sound waves have been reverberated from depths between 300 and 1500 feet, and it is suspected that the shifting interference layer causing the reverberation could be schools of fishes following the migrating zooplankton.

The zooplankton includes everything from protozoans to chordates. Chief among them are the copepods, no larger than the period at the end of this sentence. Like crabs and lobsters, they are crustaceans. Other small crustaceans are ostracods and amphipods. Foraminiferans and radiolarians are planktonic protozoans with shells that, like the walls of diatoms, resist destruction and accumulate as bottom oozes. Many pelagic and benthic animals have planktonic immature stages that aid in their dispersal. Among them are annelids, molluscs, echinoderms, lobsters, shrimp, tunicates, and fishes. An assortment of other planktonic forms includes comb jellies, medusae, Portuguese men-of-war, and arrow worms.

The quantity of plankton fluctuates widely, depending on nutrients, temperature, light, and turbulence. The effects of the fluctuations are generally transmitted link by link in the food chain from plankton to nekton (free-swimming animals), the latter mingling with the plankton in search of their favorite foods. Some of the fishes (e.g., menhaden, herring, and mackerel) have gill rakers so close together that they can strain out the tiny diatoms, thus going to the beginning of the food chain for much of their food. Baleen whales also strain out small organisms (called krill) for food by passing large volumes of water through curtainlike strainers of baleen (whalebone). From time to time the accumulated plankton is licked off the surface of the baleen and swallowed. As in land communities, predators terminate the food chain. Here the larger predators are the cephalopods, nektonic fishes, birds, and mammals.

The nektonic animals occupy all depths of the ocean. Most of them have neckless, streamlined bodies needed for rapid forward movement. Doubtless this is an evolutionary adaptation to reduce the resistance of water. Although water offers much more resistance to movement than air, it provides buoyancy so that animal skeletons do not have to be so bulky. This is illustrated by the whale, which despite its strength and mobility in water is utterly helpless when it runs aground. No land animal could possibly attain the size of the record blue whale, which weighed 170 tons, and be able to support or move its mass. By contrast, the largest living land animals are the elephants, which may sometimes weigh five tons. (The largest land animal that ever existed was *Brachiosaurus*, a dinosaur of the Mesozoic Era, which may have weighed as much as fifty tons. It, however, was amphibious, spending most of its time in swamps where its weight was partly supported by water.)

Different species of swimming animals maintain their own operating depths by means of movement, air bladders, or layers of body fat. Their vertical ranges are not well known; perhaps the widest ranging one is the sperm whale, which plunges to depths of 3000 feet and more. Being a lung breather, the whale is unable to remain below water more than forty minutes. Some kinds of whales are reported to have stayed submerged for more than an hour, whereas seals and sea lions, also lung breathers, can remain under water only about twenty minutes.

The nektonic animals are molluscs, fishes, turtles, snakes, birds, and mammals. The fishes by far exceed all the others in kinds and numbers. There are thousands of different species, exemplified by such familiar ones as tuna, flying fish, cod, herring, mackerel, and shark. Nektonic molluscs are of the cephalopod class: squids, cuttlefish, and octopuses. Representative birds are penguins, petrels, albatrosses, and shearwaters. The mammals are whales, sea otters, dolphins (including porpoises), sea cows (including dugongs and manatees), seals (including sea lions), and walruses. Some of the animals enumerated here are only part-time nektonic, spending the remainder of their time in terrestrial habitats. Even the whale, for all its fishlike appearance, shows its terrestrial lineage in its lungs.

Nekton is most abundant near the surface where the food is concentrated and there is light. Organisms involved in photosynthetic activity do not live below about 300 feet, although visibility ends at 1500 feet. Photosynthesis is best in the top 100 feet. Below 300 feet the light is dim and of short duration, producing a kind of twilight zone.

The depth to which light travels below the surface depends on its angle of incidence, which in turn is related to time of day, season, and latitude. The higher the sun is overhead, the deeper the light will penetrate (Fig. 9–1). The day becomes progressively shorter as depth increases until there is no light at all. At the forty-fifth parallel north, the length of a day in June is around 15 hours at 65 feet, 7.3 hours at 100 feet, and 30 minutes at 131 feet. Obviously, periods of sunlight are brief even a short distance below the surface.

Since fishes are visible near the surface and have no place to hide from their enemies, they depend on movement and camouflage for protection. They have evolved light colors that blend with the sky and water. Viewed from above, they are blue or green like the water beneath them. From below, their silvery or whitish bellies blend with the sky above.

APHOTIC STRATUM

Below 1500 feet, where there is perpetual darkness, the fishes have not evolved protective coloration because there is little or no selec-

FIGURE 9–1 *Day length in the ocean. At 45° latitude the day length in June varies from sixteen hours at the surface to about thirty minutes at 131 feet. (From* Principles of Ecology, *by Woodbury. Copyright ©1954, The Blakiston Company, Inc. Used with permission of McGraw-Hill Book Company.)*

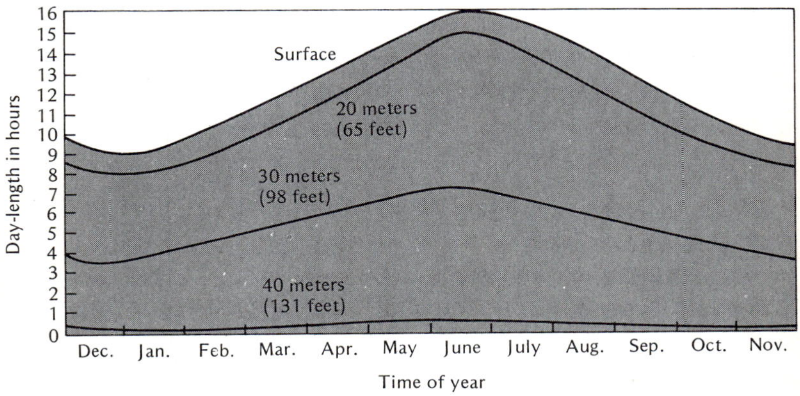

tive advantage to doing so. They occur in a variety of colors: silvery, red, brown, and black. The dark colors would be conspicuous near the surface, but at the depth where these fishes live they are invisible except by the dim and usually intermittent bioluminescence generated by scattered organisms.

Perhaps 95 percent or more of the animals living in deep water either are luminescent or harbor luminescent bacteria on their bodies. This trait is not exclusive to deep-sea life, but it is by far more common than in organisms living near the surface. Fishes with special luminescent organs swim about flashing their lights at will. Typically they have large, sensitive eyes that make the most of the short flashes. The angler fishes use flashing lights to lure their prey directly into their mouths or into a position where they can easily catch them. (The light is either in the mouth or suspended on an appendage in front of it.) In addition to attracting food, the lights help fishes see and may serve as sexual signals. The few fish that are blind or have small eyes compensate with keen olfactory perception. Many fishes of the aphotic stratum have elongated fin rays modified into tactile organs, a useful feature in this kind of habitat. With these special or exaggerated features, they are rather bizarre in appearance.

Deep-sea fishes are much smaller (a few inches long) than those at the top of the ocean and are relatively sparse. No doubt this is related to the scarcity of food, for all they can get is the fallout of dead organisms from above or a few neighboring organisms that chance to cross their paths. They have large mouths and needlelike teeth that help them catch food where it is hard to capture.

Only fishes and invertebrates live in the deep waters, never reptiles, birds, or mammals. Radiolarians, jellyfish, comb jellies, nemertians, arrow worms, squids, copepods, ostracods, amphipods, and deep-sea shrimp are among the invertebrates known to be found there as well.

BENTHOS

The benthos of the deep ocean has all the surface irregularities of jagged peaks, deep trenches, extensive plains, and slopes. The deep oceanic basin, whose boundaries are the **continental slopes**, plunges two miles or more below the surface. The overburden of water on all but the upper slopes makes the environment different in several respects. Pressure increases one atmosphere for every thirty-three feet of descent. At that rate the pressure is more than a ton per square inch at a depth of about 4500 feet compared to 14.7 pounds per square inch at the surface. At this depth oxygen is scarce. Relatively few species are adapted to such extreme conditions, and their populations are small and scattered, decreasing as the depth below the oceanic surface increases. Like much of the life in the deep water above the bottom, a high percentage of benthic life is luminescent.

The continental slopes seem to be scarred by deep gorges and canyons cut by avalanches of silt not unlike the mudflows that sometimes develop on terrestrial slopes during heavy rains. The alluvium of sand, silts, and clays spreads out on the floor in sweeping fans. Life on the continental slopes is more abundant than on the deeper ocean floors, but neither location has a flora or dense fauna like the shallow benthos close to land. Slope fauna include sponges, brachiopods, bivalves, starfishes, brittle stars, and sea cucumbers.

Besides the spill from the continental slopes, the ocean floors have picked up dust of volcanic (often submarine) eolian, and meteoritic (probably infinitesimal in quantity) origin, plus the remains of marine organisms that continually rain down from above. It is estimated that the accumulation amounts to about one inch every 2500 years. Our knowledge of sedimentation processes and distribution of deposits is fragmentary, but it is known that many of them accumulate as oozes, not yet compacted or consolidated. Named after their major components, oozes are **red** (from dust), **diatomaceous** (from silicious walls of diatoms), **foraminiferal** (calcareous shells of foraminiferans; if of the genus *Globigerina*, it is called globigerina ooze), **radiolarian** (silicious skeletal material of radiolarians), and **pteropodial** (calcareous shells of pteropods).

To keep from sinking into the ooze, benthic animals often have flattened bodies and long spidery legs that distribute their weight over more surface area, while others rise on stalks above the surface of the ooze or burrow into it. They include glass sponges, sea anemones, sea pens (hydroids), corals, sea lilies (crinoids), sea cucumbers, polychaetes, bryozoans, long-legged crabs, and tunicates.

REFERENCES AND FURTHER READINGS

Engel, L. 1961. *The Sea.* Time-Life, New York.

Marx, W. 1967. *The Frail Ocean.* Ballantine, New York.

Nybakken, J. W. 1971. *Readings in Marine Ecology.* Harper & Row, New York.

Odum, E. P. 1971. *Fundamentals of Ecology.* 3d ed. Saunders, Philadelphia.

Thorson, G. 1971. *Life in the Sea.* McGraw-Hill, New York.

Part 3
Man's Impact on the Ecosystem

Chapter 10
Productivity

THE MEANING OF PRODUCTIVITY

Organisms survive because they are successful gatherers, converters, and storers of energy. They are not, however, equally efficient in their performance of these functions. They differ as much in their ability to synthesize, use, and store food as they do in other respects.

The yardstick for measuring energy accumulation is **productivity**. Ecologists use this term in a specific sense to mean the rate of increase in organic matter, not just the amount. Thus, they are actually measuring the amount of gain in a specific period. Moreover, any calculation of productivity must also take into account the area involved. Productivity then might be expressed as the number of calories produced on a square meter in a year's time, or in terms of other units such as the number of grams produced per square meter per day.

Productivity is a function of **producers**, **consumers**, and **decomposers**. The first are either **primary** or **secondary** (tertiary, etc.) **producers**, depending on whether they synthesize food (autotrophs) or obtain it secondhand and assimilate it (heterotrophs). This means the secondary (or tertiary, etc.) producers are in reality the same as the consumers or decomposers because they depend on other organisms for food. Similarly, **primary productivity** is the rate of increase in organic matter (potential energy accumulation) produced by photosynthetic and chemosynthetic plants, and **secondary productivity** is the rate of increase in organic matter in consumers and decomposers. In any type of productivity there is always an operational (maintenance) loss, but it is much greater in animals, which are secondary producers.

Since the greatest source of energy for living organisms comes from light, the green plants are the major primary producers. They capture the light as they synthesize food. The total amount of organic matter synthesized in a given period of time—actually a measurement of rate—is customarily called **gross primary productivity**, the gain plus that used in respiration.

As consumers, animals depend on either plants or other animals for food. However, they often waste as much as they eat by selecting a portion and discarding the rest. Of all that is ingested, another portion passes through the body either unused or partially used. It is said that a grasshopper, for instance, utilizes only about 30 percent of the vegetable matter that passes through its alimentary canal. By contrast, some rodents are known to use nearly 90 percent. In addition to the unused products that pass through the digestive tract, some residue of usable matter remains in certain metabolites, such as the nitrogenous wastes that are excreted from the body but are potentially useful to some other organism. Of the amount of food assimilated, the secondary producers expend great quantities to energize muscular action.

The first requisite in determining productivity is the measurement of organic increase, a sometimes difficult and complicated process. Accurate results would require taking into account not only the amount consumed by respiration but several other losses as well. In addition, the gain must be figured on the basis of specific time intervals. The rate would by no means be steady throughout the period. Productivity is easy to measure in animals because they can be weighed at the beginning and end of a certain period. The same is true for annual plants, which can be harvested and weighed at any time after they start growing, although they are usually harvested at the end of the growing season. Despite the fact that all the growth could have occurred in a few weeks or months, the amount harvested is usually considered the annual gain. Communities of mixed perennial plants cannot be measured by the harvest method because there is no way to measure the amount of matter in the plants at the beginning of the period without killing them.

Productivity is a result of the metabolic capability of the organism and environmental variables. Its capability is genetically determined by the evolutionary process. Full productive potential is for all practical purposes impossible to achieve because the many environmental factors are not continuously optimal. That in itself imposes limitations on the number of individuals that can share whatever resources are available and at the same time maintain or exceed survival requirements. Competition, then, is a major factor in causing shortages of most necessities. Add to that the drain caused by parasites or herbivores, and most organisms fall short of their genetic capabilities.

Beyond the productivity of individuals, ecologists are often interested in the productivity of groups of individuals in specific locations (Fig. 10–1) or stages of succession. How, for instance, does the ocean compare with land environments in productivity? Which produces more, a rainforest or a coastal marsh? Is a climax community more or less productive than an earlier stage? Ecologists want to know not only which group is most productive but also how much more productive it is than others.

PRODUCTION

Sometimes productivity may be stated as the *amount* of accumulated organic matter instead of the *rate* of accumulation. To distinguish between the two, the term **production** is used for the amount of accumulation. Even then, a time period may be stated or implied. For example, saying that the production of corn is a certain number of bushels per acre implies that this amount was produced during one growing season.

When someone speaks of production as the amount of final product, he is not referring to efficiency of anabolic processes or to losses

FIGURE 10–1 *Comparison of the gross primary productivity of different habitats. Values are given as kilocalories per square meter per day. (From E. J. Kormondy,* Concepts of Ecology, *© 1969, Prentice-Hall, Inc., Englewood Cliffs, N.J. Adapted from E. P. Odum,* Fundamentals of *Ecology, 3d ed., Saunders, Philadelphia.)*

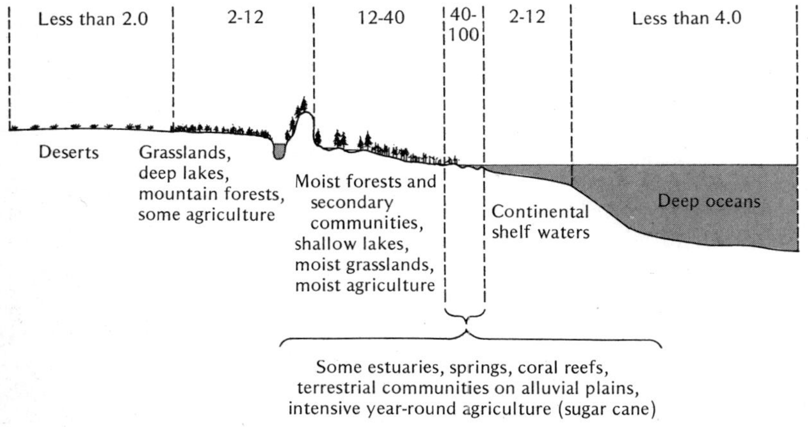

from catabolic processes, although both definitely determine the final quantity. Rather, he means the net accumulation of the final product.

The harvester and the ecologist may, however, have something entirely different in mind. The harvester is thinking about the yield (crop or harvest) of a usable product. If he is a farmer, he probably means a specific product like grain (Fig. 10–2) or potatoes, which is only a portion of the total production of the plants. If he is a forester, he means marketable timber. If he is a fisherman, he probably means the number or weight of individuals taken, which is only a fraction of the population present. The ecologist has a larger view of production—he means the total organic matter of an organism(s) or the total organic matter of all organisms in an area. He usually expresses production not as yield but as **biomass** (standing crop).

CROP YIELDS

To provide more food for an ever-increasing population, agricultural science has vigorously applied the tools of genetics, soil science (pedology), physiology, hydrology, and machine technology to increase yields. How much further yields can be increased is an open question, but the ultimate in production is not in sight. In tribute to

FIGURE 10–2 *Wheatfield in Kansas.*

the remarkable progress that has brought a surplus of food to many countries of the world and especially the United States, the developments have been fittingly termed the "green revolution."

Comparative figures show that different crops are quite variable in their yields even when grown under similar conditions. Compared to wheat, potatoes produce almost twice as many usable calories per acre; sugar beets, about three times as many. At least part of the difference can be attributed to the synthesis of a larger portion of the carbohydrates to starch and sugar, instead of cellulose as in wheat straw. For a number of reasons no one would advocate growing only the highest producer, but something might be done through genetic manipulation to increase the yields of the low producers.

Aside from improving yields through the synthesis of more usable foods, knowledge of genetics can be used in other important ways to increase productivity. Resistance to disease and drought have been bred into plants for a long time. A more recent and highly successful trend has been to produce crops that stand up better in the fields, thereby reducing spoilage. Nutritional factors, responses to fertilizers, and speed of maturation have all been altered in accordance with certain objectives. For almost any purpose desired, geneticists have a reasonable chance of success; this is true for animals as well as plants.

Animal crops have been improved by every means from genetics to food supplements. What and when to feed and the proper age for harvest have all come under scrutiny. The tendency is to turn animals into food factories by keeping them inactively confined and giving them unlimited amounts of food. The cattle feedlots throughout the Middle West (Fig. 10–3) are good examples. Birds (Fig. 10–4) are similarly confined. Laying hens are confined to very small cages in virtual egg factories. A manager of such a place has said that the space per hen has been reduced to a bare minimum—any further reduction would cause "derangement" of the hens.

Many animals generally classified as wild are still harvested where they can be found. Often the yield of wild animals can be increased if they are grown in a state of semidomestication in which some conditions can be controlled. Some terrestrial animals handled this way are wild game and honey bees. The bees (Fig. 10–5) are kept not only for honey but as pollinators of several crop plants. In some locations their hives are moved from field to field as different crops come into flower. In addition, a number of aquatic organisms like fishes, oysters, and seaweeds are managed by man in their natural habitats. The Japanese in particular have produced pearl oysters and some seaweeds for years. Fish "farming" has been practiced for centuries in both Asia and Europe, where the fishes are

FIGURE 10–3 *A cattle feedlot in Nebraska.*

FIGURE 10–4 *A turkey farm near Wodena, Minnesota.*

FIGURE 10–5 *Beehives in south Georgia. Hives are placed where nectar is readily available from flowering palmettos.*

kept in backyard pools or rice paddies as a ready source of food. Thus, without refrigeration, people can keep the fishes alive and fresh until they are needed. (Thomas Jefferson maintained such a pool at Monticello.) Carp has long been a favorite semidomesticated fish because it is a scavenger that will eat almost anything. At present several freshwater fishes are commercially produced in this country. One farmer with twenty acres of lakes supporting catfish reports a harvest of 1650 pounds per acre at a profit of about $240 per acre.

The oceans are so large and unwieldy that nothing has been done to alter their natural character except in certain locations. Since marine life will be needed to help feed a hungry world, **mariculture** has a promising future, especially in the more sheltered estuarine waters, marshes, and mangrove swamps where relatively small environmental modifications can be made to produce ideal growing conditions for a few selected species. Oysters are among the species that adapt most readily to cultural techniques. Man-made racks or even old automobile tires can be used as a base on which oysters can attach themselves. Artificial reefs of junked automobiles can also be used as shelters for some fish species. Thus far, however, little has been done beyond providing favorable environments.

Sooner or later new or improved mariculture practices will be developed. Water enrichment (fertilization) to accelerate the growth of fish foods has already been tried in some lochs in Scotland, and transplantation of fishes to new locations has been tried in many places. An example is shad, which has done well around the mouth of the Sacramento River in California. Fish breeding and predator control are other promising practices worth investigating. Finally, regardless of whether fishing is done in natural or improved waters, more effort is needed to restrain excessive fishing in some localities and to encourage more fishing in other areas.

BIOMASS

Unlike yield, biomass is the total matter (potential energy) of an individual or group of organisms. It can be expressed as weight or volume but is most often expressed as dry weight. The determination of dry weight is better than that of fresh weight because it gives more accurate comparisons. The trouble with fresh weight is that the water content of organisms varies considerably between different times of the day or year and between species. Besides, ecologists are interested primarily in the organic constituents.

To obtain dry weight the drying may be partial or complete, depending on how accurate the determination needs to be. Air or oven drying is frequently used to prepare vegetative matter for weighing. Plants are usually air dried quickly to kill them and stop any losses from respiration. If oven drying is preferred, the plants are dried at a standard elevated temperature until there is no further loss of weight.

To determine biomass one should measure total mass; however, this is virtually impossible when working with rooted plants because roots cannot be removed in their entirety. The unmeasured root systems vary a great deal in mass but can far exceed the shoot system. Because of the problem of harvesting roots, it is customary to detach the plant at the ground surface and determine only the mass of the above-ground parts. This method is particularly suited for work with herbaceous plants, which are easily clipped with garden shears. With such plants the volume of material is manageable for handling and drying. The measurement of large woody plants, on the other hand, presents a real challenge.

Like the plants, animals used to be dried in ovens, generally at temperatures of 105° to 140°F, and sometimes in a partial vacuum. Temperatures above 140°F are undesirable because the fats volatilize. The newer freeze-dry method, currently favored over drying, is an especially good technique.

The quantity of biomass can be used to compare the accumulated organic matter of different areas; to compare the relative amounts of organic matter in different habitat components such as aerial parts, roots, and litter; or, in combination with time, to determine productivity. Often the biomass of a particular area is called the **standing crop**. Some zoologists measure biomass to determine the efficiency of an organism in the utilization of food. They can easily measure food intake and what portion of it is used for one thing or another. For instance, a herpetologist studying a snake would measure gain in weight and losses through excretion, egestion, reproduction, and shedding of skin. As expected, organisms are very different in the efficiency with which they produce biomass.

MEASUREMENTS OF PRODUCTIVITY

With all of its limitations, the quantity of biomass is still used to calculate productivity, especially for certain organisms that are easily handled by this method. In addition to animals and herbaceous plants, populations of algae can be handled this way.

Under some conditions measurement of the photosynthetic rate might give a fair indication of the amount of primary productivity. Such measurements are easier to make in aquatic environments but have been accomplished in a large enclosed section of a tropical rainforest. While different techniques may be used according to their adaptability to a particular organism or community, the amount of photosynthesis is calculated from measurements of either uptake of carbon dioxide or evolution of oxygen. There are problems connected with this method because the rate of carbon dioxide uptake or oxygen evolution varies considerably as environmental conditions change. The method therefore has value only if the prevailing conditions are known and documented.

Studies have shown that plants differ considerably in their efficiency as indicated by utilization of solar energy and the amount of energy dissipated during respiration. The classic study of a cornfield by Transeau in the 1920s showed that only 1.6 percent of the sunlight was incorporated into the plants. After eliminating the amount of solar radiation that is not effective in photosynthesis (particularly wavelengths in the green portion of the spectrum), the efficiency of light utilization was closer to 8 percent. Compared to the total incident light incorporated in the cultivated corn plants (1.6 percent), the efficiency in natural habitats generally seems considerably lower. For example, a study by Guday of Lake Mendota, Wisconsin, showed the autotrophic efficiency to be only about 25 percent of that shown by Transeau's cornfield.

Net productivity may be drastically reduced by the amount of energy required for maintenance. An estimate based on a limited number of studies shows that the respiratory loss is around one-third of the gross productivity; in other words, the net productivity is almost two-thirds of the total energy captured. By comparison, it appears that trees in tropical forests may lose three-fourths of their gross productivity through respiration. For comparisons of gross primary productivity see Fig. 10–1.

Another way of measuring the productivity of primary producers (photosynthetic organisms) is to extract and measure the total amount of chlorophyll. This provides a rough basis for comparing the productivity of different communities. Although it is assumed that there is a close relationship between the amount of chlorophyll present and the rate of photosynthesis, it is also recognized that the results are subject to distortions because of the many variables that enter into the photosynthetic process.

More recently a radioactive tracer has been used to measure the rate of photosynthesis. Tagged carbon (^{14}C) is the element used. After the carbon is incorporated with the protoplasm, the radioactivity of samples (whole organisms or parts) or extracts of plants can be quantified either by using a scaler with a Geiger–Müller tube or by means of liquid scintillation counting. The use of tagged carbon is suitable for studying aquatic plants that can be kept in small containers.

By converting biomass into calories a better measurement of productivity is obtained than by using biomass alone. In this way only the organic matter is measured, and the water and ash content, which varies so much from organism to organism, does not distort the figures. Neither are the different energy potentials of carbohydrates, fats, and proteins lumped together equally. Relatively speaking, these have caloric values of 4.1 cal per gram for carbohydrates, about 9.5 for fats, and about 5.6 for proteins. Besides their quantitative differences in energy per gram, their proportions in organisms vary considerably, particularly with regard to the type of food stored. The cereals, for example, store primarily starches, whereas nuts contain a large percentage of the available energy as fats.

ANIMAL PRODUCTIVITY

Secondary productivity, that of animals (and decomposers), is not always easy to measure. Particularly difficult is the matter of finding out how much wild animals eat and how much of the ingested food passes through the body unused. Such information can be obtained from caged animals, but it cannot be reliably applied to animals in the wild that not only have fluctuating food supplies but have to forage for what they get. The foods they are able to find may be quite unlike the diets given to captive animals. While it is true that animals are selective about foods, some will substitute availability for preference when they get hungry. A sampling of stomach contents of wild animals usually shows a much greater variety of food than is normally obtained in captivity.

Even within the same population, individuals are quite different in productivity. In the adult population pregnant or nursing females can hardly be considered average individuals. Nor can old animals with lower metabolic rates be equated with vigorous young animals with high metabolic rates. Likewise, the proportion of different age classes within a population has a bearing on the productivity of the group.

Cyclic behavior, both diurnal and seasonal, is still another variable to be considered. No animal is equally active throughout a twenty-four-hour period. In particular, the activity of ectothermic animals is more closely synchronized with temperature changes than that of endothermic ones. Seasonal differences are great in most animals in that they involve periods of reproduction and of estivation or hibernation.

The importance of an animal species in a community cannot be judged solely from its numbers or size. Its metabolic rate, about which we need to know much more, is a very important factor. A small population of very active individuals could be more important than a large one of relatively inactive individuals.

To measure the importance of all animals in a large community of plants and animals is complex. The matter of identification alone is a monumental problem. Its magnitude can be judged, for example, from a report that 1700 species of insects were taken from a single stratum in a field in Michigan. In addition, rates of ingestion, excretion, assimilation, respiration, birth, and death, as well as population size (number or biomass), should be known for every animal taxon present. Of necessity, studies like these must be limited to one or a few species.

The energetics of a few animal species are compared in Table 10–1. No matter how different the taxa of animals, they share the characteristic of having to consume considerably more than they convert into biomass.

In any community the populations of secondary producers are limited to the ability of the primary producers to support them. The ability to support animal populations is termed **carrying capacity**. This yardstick of productivity is used particularly in wildlife management. Obviously, a population cannot outrun the carrying ca-

TABLE 10–1 *Population energy data from N. American old field systems and African grasslands*[a]

(All figures are in kcal/m^2)

Animals	Location	Production	Consumption	P/C
orthoptera	S. Carolina	4.00	76.9	0.05
	Michigan	0.51	3.71	0.14
	Salt Marsh	11.00	107.00	0.10
plant hoppers	Salt Marsh	70.00	413.00	0.17
	Michigan	0.08	1.51	0.05
other insects	S. Carolina	0.40	7.70	0.05
	Michigan	0.10	0.71	0.14
harvest ant	S. Carolina	0.09	34.5	0.003[b]
sparrows	S. Carolina	0.04	4.00	0.01
	Michigan	0.05	2.55	0.02
old field mouse	S. Carolina	0.12	7.40	0.02
deer mice	Michigan	0.01	1.08	0.01
ground squirrels	Michigan	0.11	5.60	0.02
ungulates	Tanganyika	4.00	273.30	0.01
elephants	Tanganyika	0.34	71.60	0.05

[a]From K. Petrusewicz, and A. Macfadyen, 1970, *Productivity of Terrestrial Animals*, IPB Handbook No. 13, Blackwell Scientific Publications, Oxford.
[b]This figure is probably unreliable and has been challenged.

pacity of the community without some of its members suffering and usually dying. This fact is no less applicable to the human population, which may soon exceed the carrying capacity of the earth; this condition is known as overpopulation.

PRODUCTIVITY OF COMMUNITIES

Most measurements of productivity are of groups of organisms rather than of individuals. For instance, an aquatic ecologist would be interested in the productivity of lake communities, but the size of such communities precludes anything but sampling. A number of sampling techniques are available for specific purposes and places, but they are usually alike in that they *count* and/or *measure*. The validity of the particular technique used depends at least in part on the size of the sample, the appropriateness of the technique to the organism, and the skill of the sampler.

If the biomass of an area is being sampled, the measuring may be done either by weight or by volume the same way one would evaluate an individual or an entire community. The biomass of most terrestrial communities is expressed in weight per unit area, usually in grams per square meter. The biomass in soil or aquatic communities is often better expressed as weight or volume per unit volume.

STABLE AND UNSTABLE COMMUNITIES WITHIN THE BIOME

As already indicated (Chapter 3), many communities within a biome are transient and succeed one another until a climax community is reached. Certain developmental trends seem to prevail from the beginning of the succession to the climax. At first the number of species and the niches available to them are few, but thereafter they steadily increase in number. Although it is generally believed that the greatest diversity is at the climax, it is worth noting that some ecologists think it may occur earlier. Intracommunity relationships among the few pioneers are simple compared to those in the more complex later stages.

Corresponding to changes in community composition during succession are the changes that occur in productivity. Generally in the early stages more energy is accumulated by photosynthesis (primary productivity) than is used in respiration. The consequence is that net productivity is high and biomass is increasing. Blooms are extreme expressions of this condition. A single-crop community (monoculture), as maintained by farming, is in reality an artificially preserved early stage in which the photosynthesis–respiration (P/R) ratio is highest—that is, net productivity is high. The climax community has the greatest amount of biomass but expends essentially all its productivity on maintenance. The P/R ratio then approximates 1.0—that is, there is no net productivity.

The relationship of production, respiration, and biomass is illustrated in Figure 10–6. The curves for a laboratory **microcosm** (a small ecosystem, in this case a small aquatic one) are surprisingly similar to those for a forest. The great difference between the two is that the microcosm reaches its climax in days, whereas the forest takes years.

COMPARING PRODUCTIVITY OF COMMUNITIES

Comparisons of the productivity of different communities must be based on limited information about variable groups of organisms

FIGURE 10–6 *Comparison of the energetics of succession in a forest and in laboratory microcosm.* P_G *= gross production;* P_N *= net production;* R *= total community respiration;* B *= total biomass. (From E. P. Odum, "The Strategy of Ecosystem Development,"* Science *164, (April 18, 1969): 262–270. Copyright 1969 by the American Association for the Advancement of Science.)*

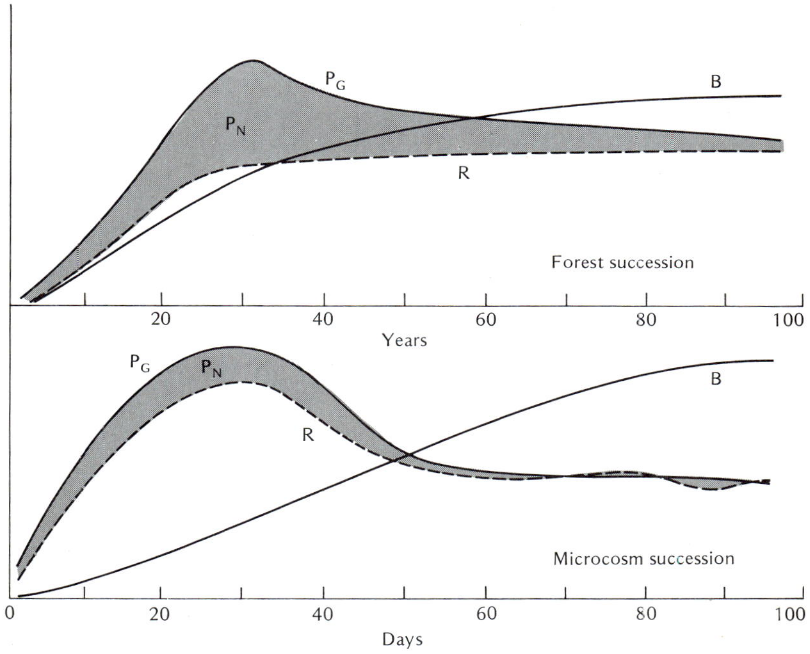

under inconstant circumstances. The conclusions, therefore, are of necessity either generalizations or specific citations of unduplicated studies.

Aquatic plant communities rank from least productive to most productive as follows: (1) marine plankton, (2) lake plankton, (3) freshwater submerged, (4) marine water submerged, (5) emergent (freshwater and marine). The primary productivity of the ocean or deep lakes is extremely low if the entire volume of the body of water is considered instead of just the upper zone where the phytoplankton lives. On the other hand, plants along the margins (emergent or submerged) of large bodies of water are sometimes disproportionately highly productive in local situations where enriching pollutants arrive from terrestrial sources.

The productivity of the ocean as a whole is low (Fig. 10–1). It produces about 50 grams of carbon per square meter annually. Its greatest productivity is in regions of upwelling where the surface is fertilized from below. Here the productivity may be as much as six times that of communities in the open ocean. Similarly, productivity along the coast may be doubled by enrichment from the land.

The productivity of a particular kind of fish is likewise related to the length of the food chain supporting it. Generally speaking, the distance between primary producers and harvested fishes is five links in the open ocean, three along the coast, and two in regions of upwelling. With a loss of about 90 percent between each link, it is obvious that a fish living in the open ocean requires a very large food base of phytoplankton to support it. The amount of phytoplanktonic biomass present at any one time, however, may be relatively small. Instead, the phytoplankton may be highly productive while experiencing a considerable turnover. That approximately half of the world's fish harvest comes from about 1 percent of the oceans, where there are upwellings, emphasizes the paucity of productivity of the open ocean.

From selected studies of terrestrial communities, it is clear that productivity varies greatly (Table 10–2). The environment is of primary importance in that it determines what and how many organisms can live there. Beyond that are all the internal factors of complex food webs and speed of population turnovers.

ENERGY FLOW

In dealing with the energetics within an ecosystem, both energy in the form of matter (potential energy) and energy in the form of work (kinetic energy) are involved. Light, the original energy source, is fixed by photosynthesis in organic matter (foods). Foods are then used by both plants and animals to build bodies and to supply energy. The transfer of energy (kinetic or potential) from sun to primary producers, from primary producers to consumers, and from primary producers and consumers to decomposers is **energy flow**.

The pathways of energy flow through **trophic levels** are short to long or simple to complex. The route taken is the so-called **food chain** (Fig. 10–7), **web** (Fig. 10–8), or **pyramid** (Fig. 10–9). The food chain is a series of organisms, each of which is food for the succeeding one. The web is more complex in that it shows numerous food relationships. The pyramid likewise indicates pathways but depicts amounts at different trophic levels.

Food chains range in length from two to several links. If it involves all trophic levels, the flow is through a chain consisting of

FIGURE 10–7 *A food chain.*

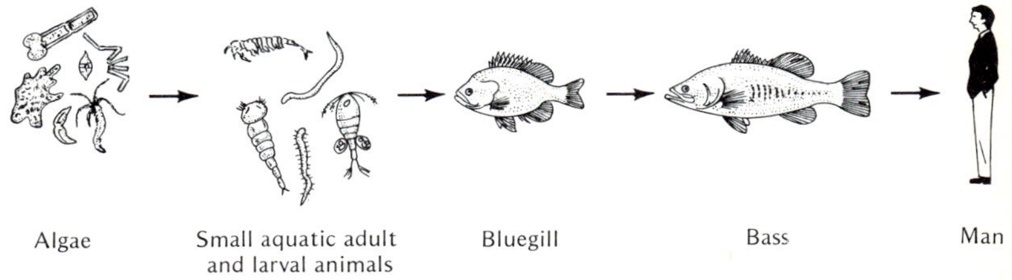

FIGURE 10–8 *A food web showing interrelationships in the prairie.*

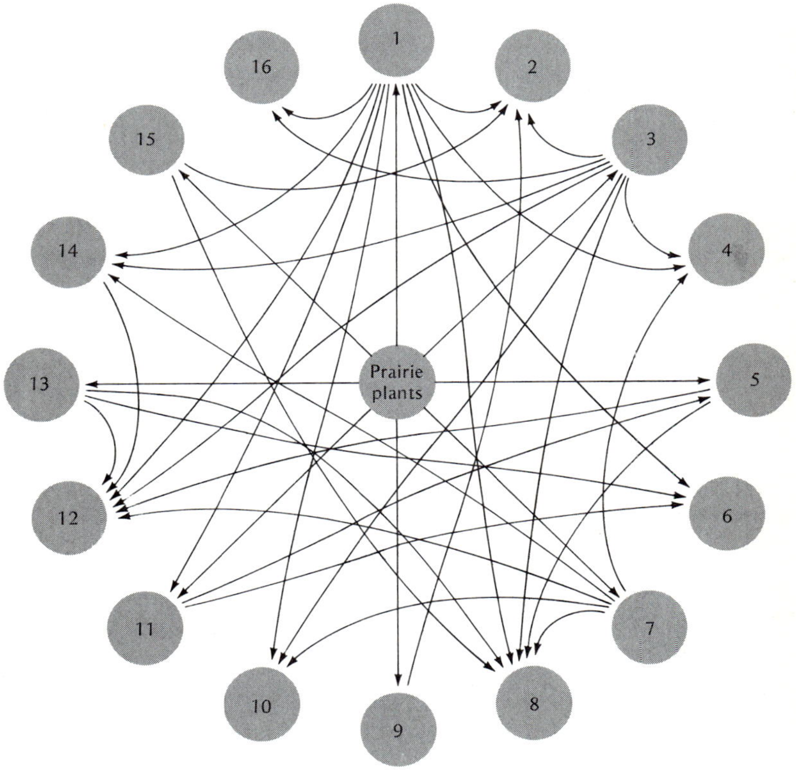

Key to organisms:
1. Deer mouse
2. Coyote
3. Kangaroo rat
4. Badger
5. Meadow lark
6. Striped skunk
7. 13-lined ground squirrel
8. Horned owl
9. Prairie dog
10. Rattlesnake
11. Grasshopper and other insects
12. Red-shouldered hawk
13. Quail
14. Bull snake
15. Jack rabbit
16. Marsh hawk

TABLE 10–2 *Net primary productivity and energy fixation estimate for the world around 1950*[a]

Vegetation unit	Size 10^6 km^2	Net primary productivity: Range gm^{-2} y^{-1}	Net primary productivity: Approx. mean	Net primary productivity: Total for area 10^9 t	Combustion value kcal/g	Annual energy fixation: 10^6 cal/m^2	Annual energy fixation: 10^{18} cal per area (col. 2)
(1)	(2)	(3)	(4)	(5)	(6)	(7)	(8)
continental							
forest	50.0		1290	64.5			277.0
tropical rainforest	17.0	1000–3500	2000	34.0	4.1	8.2	139.4
raingreen forest	7.5	600–3500	1500	11.3	4.2	6.3	47.2
summergreen forest	7.0	400–2500	1000	7.0	4.6	4.6	32.2
chaparral	1.5	250–1500	800	1.2	4.9	3.9	5.9
warm temperate mixed forest	5.0	600–2500	1000	5.0	4.7	4.7	23.5
boreal forest	12.0	200–1500	500	6.0	4.8	2.4	28.8
woodland	7.0	200–1000	600	4.2	4.6	2.8	19.6
dwarf and open scrub	26.0		90	2.4			10.2
tundra	8.0	100–400	140	1.1	4.5	0.6	4.8
desert scrub	18.0	10–250	70	1.3	4.5	0.3	5.4
grassland	24.0		600	15.0			60.0
tropical grassland	15.0	200–2000	700	10.5	4.0	2.8	42.0
temperate grassland	9.0	100–1500	500	4.5	4.0	2.0	18.0

desert (extreme)	24.0			1	—				0.1	
dry desert		8.5	0–10	3		—	4.5	—		0.1
ice desert		15.5	0–1	0		—	—	—		—
cultivated land	14.0			650	9.1		4.1		37.8	
fresh water	4.0		100–4000	1250	5.0			2.7	21.4	
swamp and marsh		5.0	800–4000	2000		4.0	4.2	8.4		16.8
lake and stream		2.0	100–1500	500		1.0	4.5	2.3		4.6
total continental	149.0			669	100.2				426.1	
oceanic										
reefs and estuaries	2.0		500–4000	2000	4.0		4.5	9.0	18.0	
continental shelf	26.0		200–600	350	9.3		4.5	1.6	42.6	
open ocean	332.0		2–400	125	41.5		4.9	0.6	199.2	
upwelling zones	0.4		400–600	500	0.2		4.9	2.5	1.0	
total oceanic	361.0			155	55.0				260.8	
total earth	510.0			303	155.2				686.9	

[a]The year 1950 has been chosen because all considerations about near-to-natural vegetation-type areas become increasingly invalid from that time on. From H. Lieth, 1972, "Modelling the Primary Productivity of the World," *Nature and Resources*, UNESCO 8, 2: 5–10.

FIGURE 10–9 *A trophic pyramid. The unlabeled shaded columns represent the decomposers.*

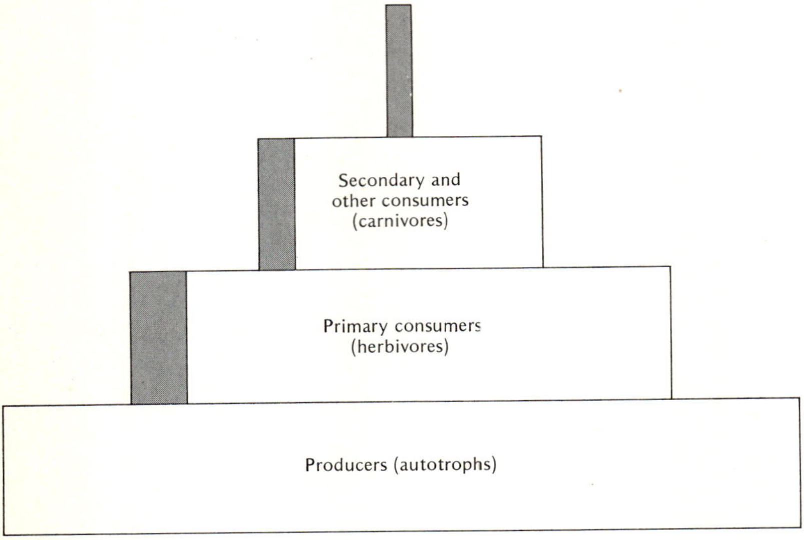

producers, herbivores, carnivores, and decomposers. The shortest sequence possible would be a two-link chain consisting of green plants and the decomposers that consume them after they die. If a herbivore like a cow consumes the plants, it is a **primary consumer**. When a human or animal predator eats meat, he is a carnivore at the third step in the chain and thus a **secondary consumer**. A carnivore might well eat another carnivore, in which case it would be a **tertiary consumer**, and so on. In all cases the final consumers are the decomposers. Figure 10–10 summarizes some terms used here and elsewhere in this chapter.

Large losses of energy occur at every step of the food chain. As already noted, the loss between successive stages is about 90 percent. The first loss of consumers is the food that remains uneaten. Of the energy (food) taken into an animal, the majority goes into maintenance, but some is used in growth and reproduction. The remainder may be lost to parasites or pass from the body as sweat, urine, feces, and gases.

Energy flow is primarily intracommunity because organisms within the community are more closely related to one another than to those outside the community. There is some leakage into and out of bordering communities as organisms cross from one com-

FIGURE 10–10 *Summary of trophic-level relationships in food chains.*

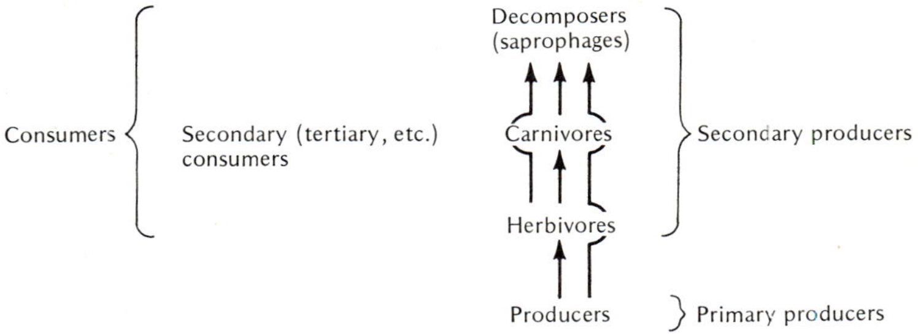

munity to the other. At the end of the food chain, decomposers are instrumental in recycling chemical components, but not energy, back into the system.

REFERENCES AND FURTHER READINGS

Boughey, A. S. 1971. *Fundamental Ecology*. Intex Educational Publishers, San Francisco.

Kormondy, E. J. 1969. *Concepts of Ecology*. Prentice-Hall, Englewood Cliffs, N.J.

Odum, E. P. 1971. *Fundamentals of Ecology*. 3d ed. Saunders, Philadelphia.

Chapter 11
Populations

POPULATIONS DEFINED

In addition to its own unique qualities, each plant or animal also displays behavioral characteristics as part of a group. Ecologists may be concerned with them as members of a mixed group, generally called a community, or as an aggregate of like individuals, known as a **population**. The latter is especially distinctive in that all of its components occupy the same niche, meaning that their requirements are identical. It naturally follows that the keenest competition is among individuals that are alike. Should some necessity be in short supply, individuals will perish. Whatever the size of the population, it is determined by "biotic potential and environmental resistance," an observation aptly stated by the ecologist Chapman.

POPULATION CURVES

Every population is limited when the supply of some basic requirement eventually falls short of its need. At that point the population ceases to grow or even decreases if reproduction outruns the ability of the environment to support it. Until that crisis point, however, populations grow rapidly according to their **biotic potential** (including their fertility characteristics and generation time). The rapid-growth phase follows a slow beginning, due at least in part to the necessity of adjusting to new locations or to the small number of individuals involved and, possibly, the difficulty of finding mates.

If the growth of a population is plotted as a curve (Fig. 11–1) from the time it occupies a new territory, it should approximate a

FIGURE 11–1 *S and J population growth curves. The J curve represents the theoretical increase due to the biotic potential of a species. The S curve represents the actual increase; this is less than the theoretical increase because of environmental resistance. Segment* a *is the* **positive acceleration phase**; *segment* b, *the* **logarithmic phase**; *segment* c, *the* **negative acceleration phase**; *segment* d, *the* **plateau phase**. *Point x marks the time when the rate of population increase slows markedly owing to environmental resistance. The broken line (plateau) represents theoretical carrying capacity; the actual population, marked by the wavy line, fluctuates near this level.*

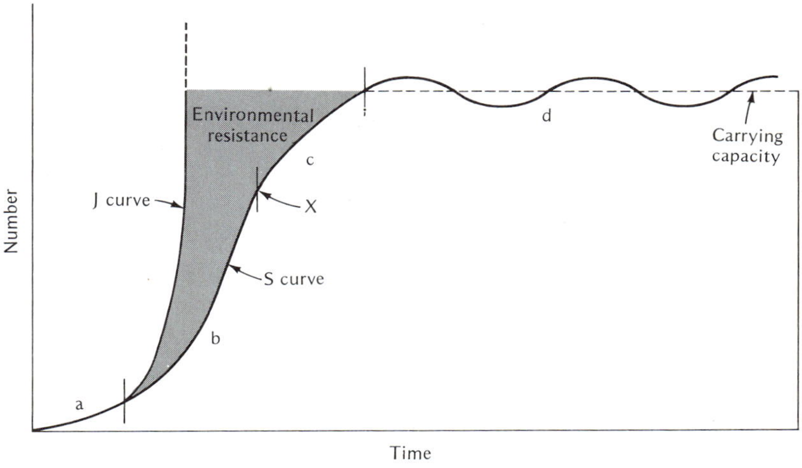

characteristic configuration often described as an **S curve**, so named because it looks like a leaning, stretched-out S. The curve may be divided into an initial gentle slope or **positive acceleration phase**, a steep **logarithmic phase**, a bending **negative acceleration phase**, and a nearly level **plateau phase**. The curve begins to bend away from the steep incline (point *x*) at the time when environmental resistance greatly increases—that is, when some factor becomes limiting. The plateau represents a state of equilibrium between the population and its environment.

A population may be limited not only by scarcity of necessities of life such as water, food, or favorable climate but also by consumers (herbivores or predators), parasites, catastrophes, antagonists, and species senescence.

The equilibrium between the population and the environment as depicted by the plateau of the curve is at best tentative. In real life it will rise and fall (Fig. 11–1), corresponding to environmental changes that upset the balance. During an epidemic, for example, the curve will dip suddenly as members die off. In an extreme case, the slaughter of the passenger pigeon by man, the curve plunged to extinction in 1914 when the last survivor died in the Cincinnati Zoo.

On the other hand, the removal of a predator may cause another logarithmic acceleration. A frequently cited example of runaway population growth is the story concerning the removal of deer predators from the Kaibab Plateau in the Grand Canyon area of Arizona. In 1906, President Theodore Roosevelt decreed that Kaibab be made a national game preserve. Hunting of deer was prohibited and their predators systematically eliminated. It is estimated that over 6000 pumas, coyotes, wolves, and bobcats were killed over the next twenty-five years. One would suspect that the removal of these predators would result in a rapid increase in the size of the deer herd.

At the beginning of the program, there were, supposedly, about 4000 deer. After the program started, their numbers definitely increased. One estimate suggests that the herd grew to 100,000 by the year 1924. Others suggest that the deer herd never increased to more than 40,000. In earlier accounts, authors claimed that the increase in the deer population was due solely to the removal of predators. But it must also be noted that over 200,000 sheep and cattle were removed from the plateau in the establishment of the game preserve. While the feeding habits of these herbivores differ, it is argued that the increase in available vegetation permitted growth of the deer population.

While there may have been some decline in the population level in later years, it most likely did not crash, as has been stated in many accounts. It seems now that the case of the Kaibab deer population has been oversimplified. In addition to predation, populations of large vertebrates are held in check by other important factors such as disease and parasite load.

Sometimes a population will accelerate in the expected way but fail to come into equilibrium with the environment as represented by the plateau of the growth curve. Circumstances may cause a die-off, which results in a precipitous drop in the curve. Since the curve up to the point of the die-off is J-shaped, it is called a **J curve**. A strict J curve is only theoretical, however, because environmental resistance results in a growth rate less than the maximum (Fig. 11–1). Therefore, it might be more accurate to call this curve an atypical S curve with a rather steep acceleration phase. This could be illustrated by plotting the growth of an algal species up to the time it produces a bloom (Fig. 11–2).

Organisms with life cycles keyed to seasonal events have J-shaped population growth curves because most of them die at the end of the season. A case in point would be a type of insect that depends

FIGURE 11–2 *A growth curve representing a population that dies off annually. The curve is in reality a series of J curves.*

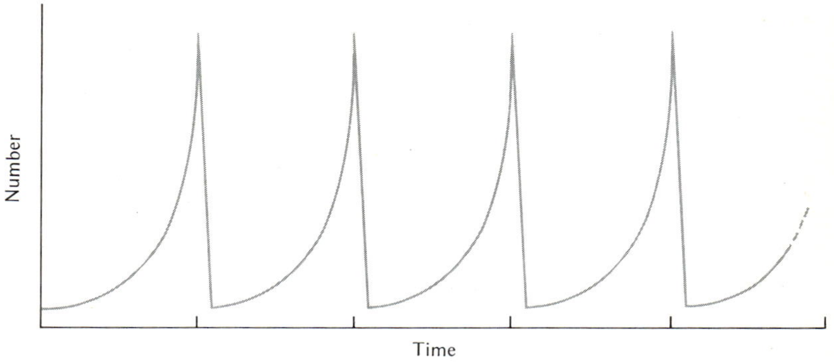

on a specific kind of annual plant for its food. As soon as the plant dies, the insects also perish. As a species the insects survive in an inactive condition, usually as eggs or immature stages, until the next season. When the plant on which the insect depends is available again, the population increases in number, following the same pattern each year.

HUMAN POPULATION CURVE

The human population curve is doubtless very much like the theoretical population curve for other organisms despite man's potential for manipulating it. One may ask: Where on the curve is mankind as of this moment? The answer depends on whether the population is viewed in a specific geographical area or as a global unit.

Population densities vary from place to place (Fig. 11–3). In some localities the population exceeds what the environment can support; this is possible only because of the inflow of food from outside sources. A breakdown in the supply system puts the population in immediate jeopardy. While populations in some localities are threatened, others are in the initial stage of expansion.

The population picture is muddled when viewed on a global basis. Some think we are reaching the tip of the logarithmic phase that surely portends disaster, whereas others feel that we are far from such a crisis. Most people would concede that we are on the logarithmic slope, and higher on it than ever before. Many are concerned about how much time is left before a limiting factor will

FIGURE 11–3 *World distribution of human population densities. (From the United States Department of Agriculture.)*

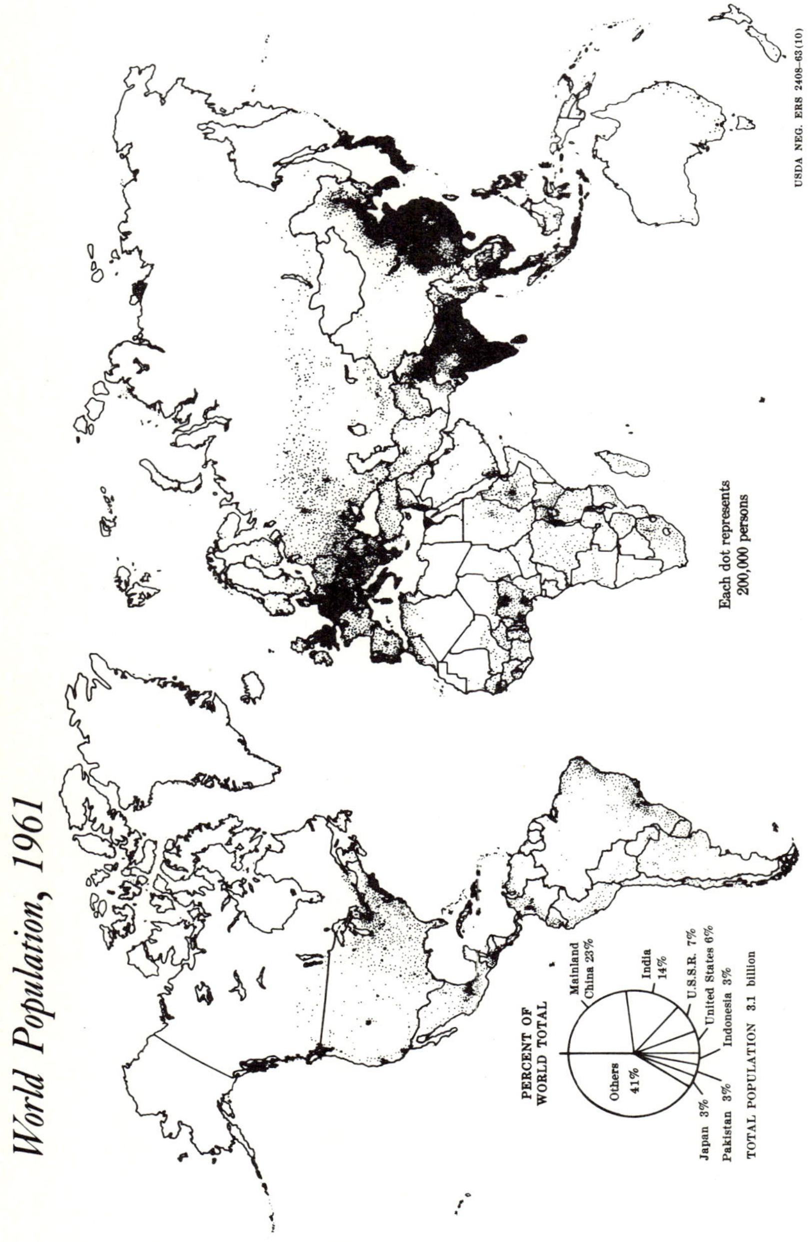

resist the upward push known as **population pressure**. If the population continues to grow, it is just a matter of time until one of the necessities of life is in shorter supply than demand. Barring such events as catastrophic war or pestilence, a shortage of food will probably be the first limiting factor, if indeed there is a single one rather than several acting together.

If it is plotted over the entire span of human history, the population curve may show a series of successively higher plateaus (Fig. 11–4). At least three major influences since the Middle Ages may have provided the potential for continued population growth and lifted the plateau to higher levels—assuming that a plateau actually existed. These influences were the discovery and colonization of new lands that provided more room to live, the medical advances that reduced mortality, and the agricultural (green) revolution that produced more food. Going further back in history, one might add the influence of man's achievement of domestication of animals and cultivation of plants, which transformed him from a nomad to a colonist.

POPULATION STABILIZATION

Although populations rise and fall in response to obvious environmental factors such as food and predators as well as a number of less obvious ones, most populations seem more stable than erratic.

FIGURE 11–4 *A stair-step growth curve showing how major historical events might have permitted several surges in human population growth.*

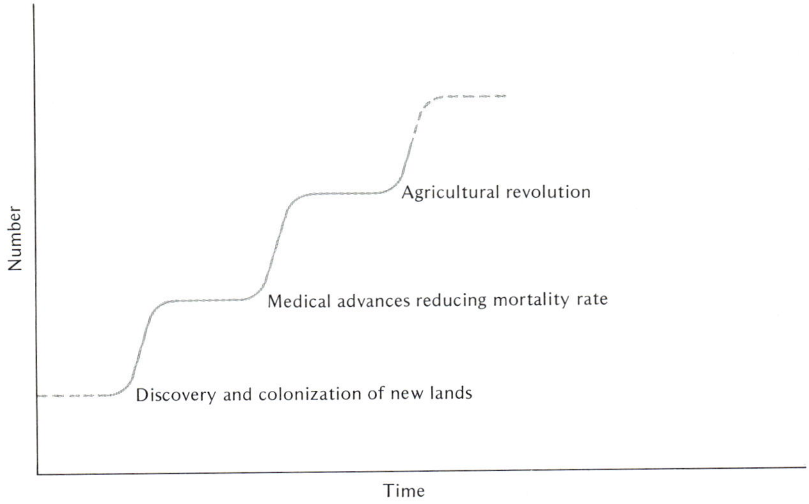

Apparently they are governed by homeostatic processes. Especially suspect is the assumption that animals will reproduce as fast as they can and that populations are regulated primarily by negative environmental controls that kill off the excess numbers.

According to one investigator, the vole (*Microtus agrestis*), a close relative of the lemming, does not seem to increase in numbers in the expected way. Instead, the population of this small rodent fluctuates rather regularly for reasons that are not clear. It does not continue to increase indefinitely, though the environment seems to be favorable enough to support larger numbers. Declines may begin quite unexpectedly. To add to the confusion, high population density does not cause an immediate decline, nor does low population density stop a decline once it has started. The cause of the majority of deaths during the population declines is unknown. Moreover, the death rate can be reduced sharply by isolating the animals in captivity.

Observations of many kinds of animals seem to indicate that population density tends to be correlated with the size of the food supply but does not normally outrun it. This means, therefore, that population declines are not normally caused by starvation. This truth is confirmed when large die-offs occur after population build-ups, despite an abundance of food. Besides, the dead animals show no signs of starvation. Apparently some subtle, self-regulating, intrinsic mechanism or behavioral routine reverses the population trend before rather than after a food shortage develops.

Much has been made of the predator–prey relationship in which the predator presumably keeps the population in check. Though cases like the Kaibab deer have been used to support this time-honored assumption, its universal or even usual application is open to question. Some now believe the predator has little or no regulating influence on a population. A case often cited in support of this view is a study made of the extent of wolf predation on caribou herds living in the North American arctic region. It was a surprise to many to find that wolves killed only a small number, mainly stragglers and weaklings whose demise actually strengthened the herd. In this light one could well conclude that the predator–prey relationship is not as much a population suppressor as a form of symbiosis.

The practice of **territoriality**, whereby many kinds of animals, notably vertebrates, select an area for their own use and defend it against intruders, is probably an effective deterrent to runaway increases in population. This is especially true of mammals and birds, which establish territories and defend them vigorously. Interestingly, their primary defense is against members of their own kind, who, because they occupy the same niche, are their most threatening competitors. In many species the territory is claimed primarily during the breeding season.

The consequences of territoriality are the dispersal of individuals and the regulation of reproduction. If space is limited, only the stronger members succeed in mating. The weaker ones that have no territories not only fail to get mates for reproduction but are more exposed to predation. They have to do without the protective advantages of a territory in which the occupant knows every nook and cranny while a would-be predator is unfamiliar with the area. Knowing all the escape routes and hiding places, the pursued animal has a decided advantage.

Among some groups, territoriality (or spacing) is maintained by cannibalism. One such case is that of a crab (*Hyas araneus*). Soft-shelled members of this species fall easy prey to hard-shelled relatives whenever they come into the range of olfactory perception.

Territoriality virtually guarantees an abundance of food for the elite members of a population because each territory is normally large enough to produce more than enough food to sustain the parents and their young. The spacing effect of territoriality must surely reduce disruptive, energy-consuming stress and infighting. In some animal groups this is accomplished by a social hierarchy or pecking order. Ample food and absence of strain contribute to good health, which in turn favors survival and stability of the population.

Another deterrent to unlimited population growth is the reaction of some animals to stress caused by living close together. Calhoun (1962) called the phenomenon social pathology, a disturbance that seriously interferes with reproduction. He observed it in a colony of eighty wild Norway rats, which he confined to a room divided into four pens. He designed the layout in such a way that the rats tended to congregate in the pens in different densities. This gave him the opportunity to observe their reactions to different population densities in each of the four pens. He found that they acted abnormally in proportion to the degree of crowding.

The densely congregated female rats built imperfect nests or none at all. Where they were most dispersed, the mothers built better nests and moved their babies one by one when they thought they were endangered. Nothing would distract them from relocating the young until the task was finished. The crowded mothers, on the other hand, would start moving the young ones but never complete the job, simply leaving them here and there throughout the pen and neglecting to nurse them. When the females came into "heat" (estrus), they were subjected to excessive attention from the males. The stress of high density had a much more telling effect on females than on males, as indicated by a higher female death rate. Many of the females died from pregnancy or birth disorders.

The male rats had their own characteristic deviations under density stress. The only males that behaved somewhat normally were the aggressive ones seeking dominance, but even they sometimes displayed peculiarities. More than 50 percent of the males abandoned the struggle for status. Some became passive, ignoring and ignored by others around them. Some became what Caldwell called "probers." They were uninterested in the status struggle but were hyperactive, even hypersexual. In their vigorous pursuit of estrous females they bypassed part of the courtship ritual. Other probers were homosexual. Many also became cannibalistic. Many of the subdominants were pansexual; that is, they were indiscriminate in their sexual targets, whether appropriate or not. They made unnatural advances to males, females not in estrus, and juveniles. Strangely, the males usually accepted such attention.

One thing is clear about the Norway rats: They cannot sustain ever-increasing population growth when they are crowded. Adult death rates were high, and infant mortality varied from 50 percent in the best conditions to 96 percent in the worst.

Like the Norway rat, other animal populations suffer under stress. Sometimes death is caused by shock as the animals expend more energy than normal in reacting to the stress. Shock follows the exhaustion of blood sugar caused by hyperactivity. In addition to deaths caused by stress, the birth rate is reduced, as has been shown in the case of muskrats crowded beyond certain limits.

That stress from high population densities restricts population growth is beyond dispute. It doubtless initiates changes in the coordinating systems, both nervous and endocrine. Since components of the coordinating systems are interrelated and delicately balanced, a stimulus that causes a reaction in any one of them may produce complicated interactions among some of the others.

Apparently the only component of the endocrine system that is directly responsive to nervous stimulation is the adrenal gland. It not only responds quickly but grows larger and secretes more in response to extreme stress. As any physiologist knows, adrenal secretions increase body defenses by releasing food reserves for energy, increasing the rate of circulation and breathing, and accelerating the coagulation of blood—all important reinforcements for fighting or protecting the individual from injury. In addition, adrenal secretions have a regulatory influence on growth and reproduction.

In the case of Sitka deer (*Cervus nippon*) on James Island, Maryland, abnormally large adrenal glands coincided with the period of greatest population density and subsequent die-off. After the die-off and the removal of stress, adrenal weight of immature deer dropped 81 percent. The coordinating systems, of which the adrenal is but a part, seemed to "break down" at the higher population level, as indicated by symptoms of shock, reduction of the stimulus to growth, loss of reproductive vigor, increased abortion, increased susceptibility to disease, and increased mortality.

External secretions as well as the internal ones just mentioned may be more important in population control than is generally supposed. Many animals communicate by means of odors (*pheromones*). Mice are known to respond by this method in several peculiar ways. When females are crowded together, their estrous cycles either stop for long periods (in most cases) or become irregular. The odor of a male initiates and regulates the estrous cycle. The presence of a strange male, however, will terminate the pregnancy of a newly impregnated female. Perhaps the odor affects the pregnant female by suppressing the secretion of a hormone needed for corpus luteum development. Without its development, estrus returns and the pregnancy is lost.

Even plants are known to produce inhibiting substances that tend to keep them evenly spaced (Fig. 11–5). They are likewise spaced by

FIGURE 11–5 *Evenly spaced desert plants on the Mojave Desert near the Charleston Mountains in Nevada. The spacing is often attributed to the secretion of growth-inhibiting substances. The larger plants are Joshua trees* (Yucca brevifolia), *which are characteristic of the Mojave Desert. (Photo by S. J. Presley.)*

competition whereby less vigorous plants perish when they cannot obtain essentials.

THE DISPERSAL OF MAN

Man is like other organisms in that he has extended his range of occupancy from a center of origin. As a relative newcomer to the planet, he has had plenty of room to live and multiply. No other species can claim such sustained and successful aggression, nor possesses his potential for traversing geographical and other barriers. Climate does not restrict him as much as it does other animals, for he is adept at controlling environments and protecting himself. He has already bridged the continents and has settled in both hospitable and inhospitable habitats. The frightening reality is that he is filling the space available, and there are no new worlds to conquer.

Wherever man has gone he has demanded much from his surroundings, often in a reckless, destructive way. If he is to continue to multiply, he must seek new worlds to conquer in the form of better ways to manage this ecosphere, not new geographical areas to exploit.

GROWING AWARENESS OF TROUBLE AHEAD

Throughout the history of man, warfare has resulted in constant and wasteful loss of energy and life. The reasons for these conflicts vary, but they often amount to struggles for limited resources. There is simply not enough of everything that may be needed or desired to go around. Reduced to the basic need for food, man has often had trouble getting it in sufficient quantity or quality.

Although recognition of the problem of food supply originated in antiquity, it was Thomas Robert Malthus, an English economist, who expressed in his *Essay on the Principle of Population* (1798) the ideas that at least set men to thinking. In essence he said that populations increase faster than their food supplies; therefore, there will always be more people than the food can support. He claimed that war and disease are necessary to kill off the surplus. Malthus' thesis that populations exceed their food supply was accepted by Charles Darwin and led him to the conclusion that only the fittest survive.

What Malthus said has not been taken as seriously as it might have been because food production has continually increased over the years. Now, people are beginning to wonder how much further man can continue to increase his food resources.

POPULATION ESTIMATES AND PREDICTIONS

The trouble with predicting the growth of the human population is that past history is no indicator of future events. Man has changed over the years, and so has his environment. There is no reason to think that the many variables affecting population size will not continue to change. There is every reason to believe that mankind is the greatest variable of them all and that he will probably do whatever is required to survive.

What frightens people today is the constant talk about the short **doubling time**, the time required to increase the population twofold. Many people believe that period is thirty-five years (2 percent a year calculated annually from the total of the previous year). Although the average annual growth rate for the world is 2 percent, the rate for individual countries may be considerably different. The current growth rate of the United States is 1 percent; Russia, 1 percent; Latin America, 2.9 percent; Africa, 2.4 percent; and Asia, 2 percent.

If the world population were close to 4 billion in 1975 and the growth rate were constant at 2 percent each year thereafter, the population would be 8 billion in 2010, 16 billion in 2045, 32 billion in 2080, etc. It is safe to say that no organisms, humans included, can maintain such a schedule. Population growth of that magnitude could indeed be called a **population explosion**. Even with a population growth of 0.5 percent, the doubling time is 140 years. Just where a crisis would occur is a mere guess. Coupled with population explosion is a phenomenon called **population implosion**. The latter refers to the increasing concentration of the peoples of the world on a relatively small portion of the earth's surface.

Every now and then one reads of the reproductive potential of various organisms and how quickly they would cover the earth if unchecked. Charles Darwin said a pair of elephants would produce 10 million descendants in 750 years. And a single bacterium, with a generation time of 20 minutes, would produce in 48 hours approximately 2.2×10^{43} individuals, or 2.4×10^{25} tons of biomass, which is roughly 4000 times the mass of the earth.

A study of human history shows that doubling times in the past were considerably longer than those projected for the future. Rough estimates of the periods required to double the population in the centuries since the birth of Christ are as follows: from 0 to 1650 (1650 years); 1650 to 1850 (200 years); 1850 to 1930 (80 years); 1930 to 1975 (45 years); and 1975 to 2010 (35 years). Each succeeding period has grown progressively shorter, which means the growth is accelerating. What are the reasons?

CAUSES OF POPULATION GROWTH

The direction of population changes is indicated by the difference between birth (natality) and death (mortality) rates. To sustain the trend of human population growth, the birth rate has obviously exceeded the death rate. Evidence also indicates that reduction of the death rate is another contributing factor to population growth.

Life expectancy tables (Table 11–1) show a steady increase in length of life. Reduction of infant mortality, conquering of disease, improved nutrition, and better living conditions help increase life expectancy. Not too long ago, medical practices were probably more harmful than helpful, more superstitious than factual. Bacteria were first seen in 1683 and first linked with disease in 1876, only about 100 years ago. Even the medical practices of the previous generation look primitive by today's standards, which include preventive measures as well as the healing art.

Consider the dismal state of affairs in the Middle Ages when the Black Death (bubonic plague) swept over much of the earth. Estimates of death from the plague in England in the year 1348 vary from one-third to one-half of the total population. During the years 1348 to 1350, at least 25 percent of the population of Europe was wiped out.

Almost everybody believed the Black Death was a punishment from God and reacted to it in strange and extreme ways. Expecting life to be short, some turned to lust and drink in an effort to extract

TABLE 11–1 *Estimated life expectancy for human populations in several locations*

Time	Location	Approximate age in years
1000 B.C.	Greece	under 20
ca. A.D. 30	Rome	23
Middle Ages	Europe	30
1776	United States	35
1850	United States	40
1900	United States	49
	India	24
1950	United States	69
	India	32
1960	United States	70
	India	42

the last bit of pleasure from life. Seeing good and bad people alike dying about them, they rationalized their behavior by claiming that both were destined to come to the same sad end. Others turned to religion with extreme zeal and sought God's mercy by giving extravagantly to the church and making hard-to-keep vows. Instead of attributing the plague to germs, which were unknown, some people accused Jews or physicians of poisoning them for selfish reasons or to help the devil. Now, however, superstition, witchcraft, and misguided practices have lost most of their followers to scientific physicians who have demonstrated their competence in healing.

Nutrition has also come a long way in recent generations as a result of better knowledge, transportation, and techniques of food preservation such as refrigeration. Americans forget that ice making, now considered a necessity, is a recent discovery and still unavailable to many people. Open-air markets where dust, flies, and filth abound are still the principal source of food in less developed countries of the world. Many foods formerly used only in season are now available throughout the year. These include vegetables in winter and meats in summer. Many countries like the United States and Mexico have greatly improved their agricultural practices, thereby producing an abundance of the high-quality food essential for good health.

Birth rates have probably been high throughout much of the history of man, with declines coming only in recent times. Early man certainly could not have understood the reproductive process very well—births must have been regular occurrences, to say the least. Children were wanted or needed in earlier societies for several reasons. From a very young age they were given the responsibility of herding animals or tilling the soil. With the coming of the industrial revolution, children also worked in the factories to help support the family. Before the day of private and government pensions, people generally wanted children as a form of security for their old age.

These reasons for having large families are rapidly losing validity as society changes. In fact, most couples could hardly justify having children from a purely economic point of view: Most children have to be supported for twenty or more years and contribute little or nothing to their own maintenance. The cost of education, in particular, is an ever-increasing burden. As soon as children can survive on their own they commonly "leave the nest," while the parents remain at home alone and unattended.

Until recently most governments encouraged a high birth rate on the assumption that there was strength in numbers. The more bodies available, the larger the labor and military pools, the greater the nation's productive capacity and international stature. U.S. tax laws still favor those who have children. Japan formerly encouraged population control as a national policy but is now concerned that if the trend is not reversed the country will soon have a disproportionate number of old people who are less productive.

CONSEQUENCES OF OVERPOPULATION

Man needs only to look at other animal populations to see the consequences of overpopulation. In addition, he has been privileged to foresee the effects of overpopulation in parts of the world where human concentrations are already at crisis levels. The storm flags are flying. Enough time may be left to prepare for the storm if the warning is taken seriously and responses are immediate.

One of the first signs of changes to come will be a lowering of the standard of living—not necessarily starvation but scarcities and perhaps rationing. Austerity will replace affluence in such matters as food, where vegetable products will gradually supplant meats and other animal products.

Along with a decline in the quality of foods, overpopulation will also affect the supply of creature comforts and conveniences like clothing, housing, heating, cooling, and transportation. Acute supply shortages will develop because of inadequate distribution or unavailability. Indeed, some resources may be completely exhausted. One of the most likely problems is a shortage of space. Already there are signs that crowded people find it increasingly difficult to channel their energies to constructive purposes, as indicated by increased indolence, restlessness, and crime. In the parlance of prisoners, humans may become "stir crazy" when they feel confined by congested conditions.

Moreover, humans cannot consume without also polluting, if for no other reason than that they eliminate body wastes. Of course pollution accompanies industrial processes, burning fuels, changing atomic structures, farming, and a host of other activities. The problem is that the whole ecosystem is altered, doubtlessly for the worse.

The security of man in the ecosystem is inseparable from the stability of the system itself. Man is already responsible for the extinction of several species—a loss that is not easily measurable because all components of the system are interrelated, often in subtle and complex ways. Even if their extinction did no harm to the system itself, these creatures are no longer present for people to see and hear. Man might get along without eagles, sequoias, or elephants, but what a loss!

Many view with alarm the probability of exhausting the earth's natural resources. That some important resources will soon be used up is expected. Although Americans amount to only about 6 percent of the world's population, they consume from one-third to one-half of the world's raw materials.

More and more attention is being directed to conservation. It may involve not just the elimination of waste but increased thriftiness as well. One possibility currently attracting attention is recycling, in which used items like bottles and tin cans are saved and either reused or broken down for their raw-material content. There are, however, optimists who contend that exhaustion of certain resources will not be critical, saying that man can and will find substitutes as they are needed.

As a biological necessity and because of its abundance in the environment, water is one of the most valuable natural resources. Its uses are too obvious to need enumerating and are of such recognized importance that a continuous supply must be maintained. Laws governing the allocation of water are stringent in arid lands where the value of water is felt most keenly. Nevertheless, even cities like New York, in climatic areas where rainfall is usually abundant, have experienced shortages.

If rainfall were distributed equally over the earth's surface, it would amount to about forty inches per year. Unfortunately, it is unequally distributed and unequally effective where it falls; much of it is wasted. Unless it is quickly used, a large portion of the water evaporates or rushes to the sea, where it becomes part of the great reservoir of salt water comprising 97 percent of all the world's water. Of the remaining 3 percent, three-fourths or more is trapped in ice. Transforming sea water into fresh water and pumping it landward against the pull of gravity is very expensive. Such problems emphasize the fact that the water supply is not limitless.

Soil, like water, is one of the world's basic resources. Its contribution to our food supply far exceeds that of the ocean. Nevertheless, it has suffered from improper use through ignorance and indifference. With the increase in demand for food to feed a growing population, the gravity of the situation is obvious.

Erosion is the most pressing problem of soil conservation. Not only is valuable soil lost but it is redeposited in reservoirs, water courses, and harbors where it interferes with other human activities or damages natural habitats. In many places forests have been all but destroyed, leaving the soil unprotected. The rim of the Mediterranean Sea and some Asiatic countries like China, India, and Pakistan are places where forest destruction has been extensive. The overcrowded populations of these regions have been forced to destroy the trees to meet increasing demands for fuel and other purposes, and replacement by new growth is running far behind destruction by the axe. The condition is so critical in places like India and Pakistan that millions of people have had to resort to the use of cow dung for fuel. The land in such places has suffered extreme damage from erosion.

Another consequence of overpopulation is loss of individual freedom. It is infinitely easier to trespass on the rights of others if one is in close proximity to them. In a manner of speaking, the individual's territory is greatly reduced. The extreme possibility is that social needs will take priority over individual needs or desires; indeed, many government policies already point in that direction. Essentially, this means forced sharing through limitation of property rights or taxation. Although the human animal seems to have a social sense, he prefers to express it voluntarily.

Finally, overpopulation increases the risk of war. The have-nots will take from the haves if conditions get bad enough for them to conclude, "What have I got to lose?"

HOW POPULATION CAN BE CONTROLLED

Fortunately, population growth can be controlled if mankind has the knowledge and the will to do something about it. The obvious first step is education. Everybody, not just leaders, needs to know the consequences of overpopulation and the alternatives. Every nation, not just the United States, needs to apply knowledge to solving the problem.

One reason for the importance of education is that it can influence attitudes about the desirable number of children per family. For purely selfish reasons, prospective parents should realize that children are economic liabilities, especially in a deprived environment. This consideration was taken seriously by victims of the Great Depression, and as a result one or two children became the norm of the 1930s. After World War II, approximately ten years later and in prosperous times, the figure doubled.

Parents are also obligated to their society for enjoying its privileges. If the ideal population level can be determined with clarity, all should willingly abide by guidelines to prevent it from going higher. Probably one of the best deterrents to large families would be public disapproval.

Once the optimum family size is established, it can be achieved by birth control, abortion, and family planning. Whatever birth control method is chosen, the user must understand it fully. There are literally millions of ignorant people who are uninformed about how their bodies function and unprepared to perceive their responsibility. Stories are told of women who have taken birth control pills on alternate days instead of daily and been surprised by the results. Another woman is reported to have dumped birth control, vitamin, and other pills into a little bowl on the kitchen table and taken them one at a time regardless of which kind she touched first. She, too, was surprised by what happened! Cases are also known where men have taken the birth control pills intended for their wives.

Sterilization is really a birth control method and seems particularly desirable for those who already have all the children they want. It has long been recommended for males because of the simplicity of the operation. Sterilizing the male consists of severing and sealing the sperm ducts (**vasectomy**). Females can be sterilized by a similar surgical procedure on the fallopian tubes (**salpingectomy**).

A new and drastically simplified surgical procedure for sterilizing women has recently been perfected at the North Carolina Memorial Hospital in Chapel Hill. Through a small incision in the navel, the surgeon inserts a tube (laparoscope) that serves as a tiny window through which he can observe what he has to do inside the abdominal cavity. He inserts his operating instrument, a slim biopsy forceps, through another tiny opening just below the navel. With the forceps he electrically seals a segment of each of the fallopian tubes, then divides the tubes through the sealed portions. After the patient is anesthetized, the entire operational procedure requires only about twenty minutes. The incisions are so small that only band-aids are used to cover them.

The intrauterine device is another method of birth control. By interfering with implantation it prevents the early embryo from continuing its development. This is, of course, quite different from preventing fertilization, which is often what people mean by birth control.

Should the attempt to prevent conception fail, abortion is a substitute. The most surprising development in this area is the sudden acceptance by so many people of a procedure that was so recently illegal and morally rejected. New laws regulating abortion and defining the stage of development at which the fetus should have legal rights are sure to be enacted soon.

Reducing the birth rate is just a part of the broad concept of family planning. Family planning means advanced determination of family size and its achievement through birth control, abortion, or adoption. In other words, it means controlling natality if the couple is fertile or obtaining children by adoption if they cannot have children of their own.

Many suggestions have been made for controlling the overall

population—some as drastic as infanticide, which has been practiced by a number of primitive tribes. Such extreme measures are both undesirable and unnecessary. Some practical suggestions are as follows:

1. Require sex education in schools.
2. Make birth control information, materials, and services readily available, perhaps without charge.
3. Make adoption procedures simple and economical.
4. Provide tax advantages for late marriages and small families.
5. Encourage sterilization, especially after a certain number of children.
6. Encourage public acceptance of the desirability of having small families.

REFERENCES AND FURTHER READINGS

Adams, L., ed. 1970. *Population Ecology*. Dickenson, Belmont, Calif.

Boughey, A. S. 1968. *Ecology of Populations*. Macmillan, New York.

Calhoun, J. B. 1962. "Population Density and Social Pathology." *Scientific American 206*, 2: 139–48.

Chitty, D. H. 1960. "Population Processes in the Vole and Their Relevance to General Theory." *Canadian Journal of Zoology 38*: 99–113.

Ehrlich, P. R. 1968. *The Population Bomb*. Ballantine, New York.

Hall, R. E., ed. 1970. *Abortion in a Changing World*. Vol. I. Columbia University Press, New York.

Hardin, G., ed. 1969. *Population, Evolution, and Birth Control*. Freeman, San Francisco.

Kistner, R. W. 1969. *The Pill: Facts and Fallacies About Today's Oral Contraceptives*. Delacorte Press, New York.

Krebs, C. L., 1972. *Ecology*. Harper & Row, New York.

Slobodkin, L. B. 1961. *Growth and Regulation of Animal Populations*. Holt, Rinehart & Winston, New York.

Solomon, M. E. 1969. *Population Dynamics*. St. Martin, New York.

Chapter 12
Human Food Supply

CHANGING NEEDS

It is generally believed that at least two-thirds of the human population subsists on inadequate diets. How then can twice that many people be supported thirty-five years from now when the population is expected to have doubled? Some insist that a world food crisis is already at hand.

When man was in the hunting and gathering stage of development some 10,000 years ago, he probably required up to two square miles of space per capita. His requirements were considerably reduced when he learned to tend animals and till the soil. Besides reducing his territorial needs, he considerably improved his mode of life in that he saved energy, eliminated many hazards of the search, and had a ready and dependable supply of food. Now he could hold his meat on the hoof until needed, thereby eliminating the uncertainty of the hunt and the spoilage accompanying it. Beyond that first advance, which changed his whole way of life, he has come a long way; he now requires much less space.

PRODUCTIVITY AND HUMAN NEEDS

Modern man requires only about one acre of fertile land per capita. The United States is fortunate in having 2.9 acres per capita, which means it can either produce a large surplus or increase its population considerably. Canada is unusually fortunate in having 6.5 acres per person. Unfortunately, many countries do not have as much as one acre per person and therefore are not self-sufficient. These countries must get along without the food they need or make up the deficit with imports. Conditions are aggravated during the inevitable periods of crop failure. India, with 0.8 acre per capita, experienced crop failures in 1966 and 1967; during those years the United States shipped 20 percent of its wheat crop to India. Population and productivity are even more unbalanced in China, with only 0.5 acre, and England, with only 0.4 acre per person.

The future could be quite different, depending on the rates of population growth and food production, both of which can and probably will change. At present the annual population growth rate in the United States (and Russia) is well below the world average of 2 percent. Simultaneously, the agricultural productivity growth rate is about 2.3 percent. This permits the United States to grow a large surplus for export. The surplus is so large, in fact, that the government pays farmers not to grow so much. In the future people may look back to this time with envy.

On a percentage basis, Mexico's annual growth in population and food production surpasses that of the United States. Mexico's population growth rate is 3 percent while its productivity increases 4 percent annually. Latin America and Africa, on the other hand, are losing ground rapidly in that their inhabitants are multiplying faster than they can increase the food crop to sustain them.

INCREASING YIELDS

In their efforts to increase the food supply, scientists will doubtless continue to apply genetics to the breeding of crop plants. Some plants will be bred for greater productivity. An example is the new variety of rice produced by the International Rice Institute of Los Banos, Philippines. Within the short span of five years, a new rice known as IR-8 was developed with characteristics capable of tripling or quadrupling the yield per unit area. Rice and other grains now account for approximately two-thirds of the cultivated acreage of the earth and are the primary sources of starch, on which both man and beast depend for a major part of their energy. This is one of the major accomplishments of the **green revolution**, the improvement of agricultural productivity by means of high-yielding crop plants and fertilizers.

Other grains besides rice have been improved. A notable achievement is the development of hybrid corn with superior vigor and uniformity. For future use, Mexico has assembled a "germ plasm bank" of over a thousand different varieties of corn from which characteristics can be selected as needed. This is a collection of variable but interbreeding organisms whose differences are caused by the many different genes they contain—in other words, a large collection of genes.

Plants will be produced that are sturdier, drought-resistant, disease-resistant, insect-resistant (e.g., wheat to Hessian fly, alfalfa to aphid and leafhopper, and corn to corn borer and earworm), more responsive to fertilizers, and richer in protein. Others will be less sensitive to length of day and hence adaptable to large geographical areas. Some will mature in a shorter time, making it possible to grow them in colder climates where the growing season is short or in arid regions where the rainy period is brief. Perhaps some plants will be developed that will transpire less water.

Animals will be bred that are more productive and more efficient converters of food to meat and other products. Already, domesticated hens in the United States are said to average 220 eggs per year compared to the dozen or so laid by a wild hen. Similarly, a modern cow produces about 9000 pounds of milk annually compared to the estimated 600 pounds produced by the ancestral cow. Quantitatively and qualitatively, domesticated animals are far superior to earlier types in satisfying the purposes for which they have been bred.

Along with genetic improvement of plants and animals, agricultural experts will try other ways to increase yields. Among them will be new cultivation techniques, improved fertilizers and methods of applying them, better herbicides and insecticides, soil conditioning, improved handling and storage of crops, and more and better use of irrigation.

Irrigation consumes large quantities of water. In some places as much as half the irrigation water is lost by evaporation into the air or absorption into the ground as it flows through open unlined channels on its way to the fields. The loss is greatest where the ditches are long, the ground is porous, and the stream has a large surface exposure relative to its volume. The loss can be reduced drastically by lining the channel with concrete or, better still, using pipes instead of open channels.

Of the unwasted water that reaches the fields, a large amount is required to flush natural salts from the soil before crops can flourish. Not long ago, flush water from fields along the lower Colorado River was rechanneled to the river and used again downstream on cotton fields in Mexico, where it severely damaged the crops. Twin problems encountered in other parts of the world are waterlogging, due to raising the water table, and salting near the surface, caused when the water evaporates and leaves its solutes behind. Ancient civilizations in Western America and the Middle East apparently died because of these problems. The same was happening in West Pakistan a few years ago, before farmers learned to reuse the groundwater for irrigation. By utilizing the groundwater they lowered the water table, and at the same time the water carried the salts away from the surface as it percolated downward.

Despite problems like these and others as well, small portions of deserts have been made to bloom. Even in regions where annual rainfall is high and crops grow well without irrigation, they grow even better if water is supplied during periods of inadequate rainfall (Fig. 12–1).

In some localities two or more crops can be grown consecutively throughout the year. For this practice to succeed, the climate must be mild and the crops carefully selected and scheduled. It has been unusually successful in choice sites in the state of Sonora, Mexico, where the growing season is long and ample water is available for irrigation. In northern India many farmers grow an early-maturing winter wheat followed by rice. Now that fertilizers are generally used, multiple cropping of level, good-textured soil is much more desirable than farming single crops on steep, eroded land that is economically marginal at best.

FIGURE 12–1 *A Georgia cornfield wilted by drought. Though annual rainfall may be ample, crops can suffer as a result of irregular distribution of rain.*

Many variables determine the yields of different crops and their suitability for growth in a specific location. Where different crops can be grown in the same place, some perform better than others in terms of usable calories produced. That criterion alone might be applied to crop selection if the need is urgent enough to force people to sacrifice their food preferences for something more abundant but less palatable.

In an effort to grow larger crops in arid lands, dry farming is often practiced. Sometimes farmers plant and take a chance that the season will be favorable. Others divide the fields into strips that are alternately planted and left fallow (Fig. 4–1). The fallow strips accumulate water for use during the following year, when they will be seeded. They are cultivated while idle, a practice that keeps out water-consuming weeds and reduces the loss of water brought to the surface by capillary action. When these strips are again seeded, the strips that have just produced a crop are left fallow. This kind of farming is practiced extensively on high plains, as in Montana. Some desert-dwelling Indians do patch farming by planting garden-size plots and small orchards (Fig. 12–2) at the bottom of slopes where the ground is soaked with water that drains from higher ground during the infrequent rains. Here dry farming is a borderline operation.

The longstanding practice of clearing additional land to produce more crops may be reaching a point of diminishing returns. Some swamps remain to be drained, some arid lands to be irrigated, and some hilly country to be put to grass; yet most of the best land has already been selected and utilized. The marginal unproductive land that is put to plow is often abandoned either because it is unproductive from the start or because it becomes progressively poorer as a result of erosion.

Still untapped for agriculture are large tracts of tropical forests, which ecologists claim are not as promising for farmland as the luxuriant growth of the trees would seem to indicate. The soil is actually infertile. Most of the mineral nutrients are tied up in the vegetation itself. If it were destroyed, these nutrients would be released and rapidly leached from the soil; removal of the plants would prevent the minerals from recycling in the soil.

As a result of the agricultural revolution, the small and inefficient family farms have gone the way of the horse and plow, and have been replaced with large, mechanized farms run by college-trained agricultural experts. Farming is a business requiring brains and aggressiveness on a par with the best leadership in other competitive enterprises. Hand labor is out, resulting in a lower production cost per unit.

FIGURE 12–2 *A garden and fruit trees planted by a Navaho Indian family in Canyon de Chelly, Arizona. The low ground is suitable for gardens and orchards because it receives supplemental water as runoff from higher ground.*

The agricultural revolution continues, however, and radical ideas are being tested. One is a pilot project sponsored jointly by the University of Arizona and the University of Sonora at Puerto Penasco, Sonora, Mexico. The effort is to provide three basic essentials—water, food, and power—in a single economic package. The facility is designed for use on desert coasts where salt water can be desalinated and used. The power is generated by diesel-driven generators. Waste energy from the generators operates the desalination process. Food is grown in an inflatable greenhouse where the temperature is regulated by salt water and the atmosphere is enriched with carbon dioxide from engine exhaust fumes cleaned in a seawater scrubber. Preliminary runs of the system seem promising, at least for use in areas where nothing better is available.

THE OCEAN AS A FOOD SOURCE

The ocean has often been viewed as a final, limitless source of food on which man can rely when the land can no longer support him. The truth of the matter is that most of the vast, open ocean is a biological desert (Fig. 10–1) where nutrients are scarce. This is indicated by the presence of much larger populations of plants and animals along shorelines, where enriching materials wash in from the continental watersheds, and in the zones of upwelling, where nutrients rise to the surface. Production as measured by biomass is about twice as great in the shallow waters near the shore and even more in areas of upwelling. One problem in utilizing ocean life is how to develop efficient and economical harvesting techniques for some organisms not now used by man. On the other hand, there are signs that techniques devised to capture old favorites like certain species of fishes and whales are so successful that these organisms are threatened, as indicated by decreasing yields.

HUMAN FOOD PREFERENCES

Always remaining is the hope that some new organisms possessing special advantages may be discovered or popularized as food sources. At present only a very few plants and animals out of thousands of existing species are used for food. An example of a possible new crop is the cactus. At the National School of Agriculture at Chapingo, Mexico, sixty **clones** have been included in the germ plasm bank. The geneticists there are interested in producing high-quality fruit for food and tender, spineless pads for vegetables or forage.

Man is peculiar about his selection of foods, preferring only the ones he is accustomed to eating. Within the human family there are wide differences in food choices based on cultural backgrounds. For example, some Orientals like squid or seaweeds, the French like snails, and some Pacific islanders like palolo worms—all more or less rejected by the average American. The diet of the aborigines of Australia includes many items like succulent grubs from the wood of eucalyptus trees, which most people would shun unless faced with starvation. The fact that man is unusually selective in the kinds of true fishes or shellfishes he consumes results in the rejection of tons of "trash" fishes that are either thrown away or used for something else. Such foods may not be as palatable as the ones he selects, but the time could be near when food habits will have to change.

A likely worldwide trend is the substitution of vegetable matter for meat. For many Americans this would mean going from steaks to ground meats, to ground meat mixed with fillers made from animal scraps or vegetable matter, and finally to plant products enriched with a small quantity of animal products. The shift is already a reality in some overpopulated parts of the world where poultry and fish, when available, have long since replaced grazing animals as sources of meat. Even in the affluent United States, the trend away from meats is already evident to a limited extent. Although meat products have been diluted with plant products for years, a recent change is the addition of poultry products to some brands of wieners.

As bad as it may sound, it is possible that some foods now going into the pet food market will have to be used by humans. Pet foods come from animals whose meat man considers undesirable, such as horses or kangaroos, or from parts of animals not normally eaten by him, such as kidneys, intestines, hearts, lungs, brains, glands, and bones. They are, however, good foods from a nutritional point of view.

Getting rid of religious taboos against some foods would be helpful in countries where certain sects are unusually large. Unfortunately, the people who can least afford to discriminate against certain foods are often the ones who discriminate the most. Neither can people who belong to certain religious sects afford the luxury of putting a high value on sacred cows, monkeys, or other animals at the expense of human needs.

That food preferences can shift dramatically is documented by history. Food plants taken from this country after it was discovered have been adopted elsewhere and have become major crops. Among these are sunflower (especially in Russia) and corn (worldwide). Another, the Irish potato, became so important in Ireland that its significance is indicated by its name. Brought to this country were cattle from Eurasia, soybeans from China, grain sorghum from Africa, arid land grasses from Africa, and many other plants. Surprisingly, many organisms thrive when introduced into new geographical locations. Some even replace the indigenous fauna and flora (Fig. 12–3).

FIGURE 12–3 *Kudzu* (Pueraria lobata) *covering native trees. The plant comes from eastern Asia and has been widely used in the Southeast to check erosion. Unfortunately, it is very difficult to control once started.*

SYNTHETIC FOODS

Synthetic foods, imitation foods, or food supplements could become increasingly important in the near future. They can be made from abundant, cheap, or less desirable food products. Imitation foods such as "bacon" that looks and tastes like the real thing are already on the market. A particularly promising food supplement is fish protein concentrate (FPC), an odorless, tasteless powder made from whole trash fish; the unpalatable oils are extracted and used for other purposes. Two other vegetable protein concentrates in powder form now on the market are **incaparina** (25 percent protein) and **laubina**. The former is a mixture of cottonseed and corn meals enriched with yeast, vitamin A, vitamin D, and calcium carbonate. The latter is made from wheat, chick peas, bone ash, sucrose, and some vitamins.

BETTER USE OF NATURAL ENVIRONMENTS

Almost all discussion on the subject of enlarging food supplies emphasizes genetic alteration of organisms or practices that change natural environments. From the standpoint of agriculture some natural environments are useless, especially for cultivated crops. Some lands are so marginal that they are ill-suited even to support a single type of domesticated herbivore, since all members of a herd select and reject the same kinds of plants (Fig. 12–4). Moreover, cattle in particular damage the fields with their trampling. Better utilization of land suitable for grazing might be achieved by stocking it with native wildlife. Here several species are usually involved, each filling its own particular niche but not consuming exactly the same kinds of plants in the same proportions.

Studies made in Rhodesia have shown that wild game are more efficient grazers and more productive assimilators of protein than the cattle introduced from other environments. They are likewise more adaptable to the native vegetation, scarcity of water, and local diseases. By comparative cost analysis the native game were found to be about six times more profitable, with the difference due largely to the higher cost of maintaining imported domesticated animals. With such an advantage, game ranching should become more popular.

FIGURE 12–4 *Evidence of selective grazing by cattle in a Kansas pasture. The grass selected is buffalo grass* (Buchloe dactyloides)*; the rejected forb is thistle* (Carduus). *(Photo courtesy of G. W. Tomanek.)*

TABLE 12–1 *Food chains and energy losses[a] at different trophic levels*

Trophic Level	Chain 1	Chain 2	Relative grams of food available
pentary consumer	man	—	1
	↑		↑
quaternary consumer	tuna	—	10
	↑		↑
tertiary consumer	mackerel	eagle	100
	↑	↑	↑
secondary consumer	herring	snake	1,000
	↑	↑	↑
primary consumer	zooplankton	rat	10,000
	↑	↑	↑
producer	phytoplankton	grass	100,000

[a]Assuming a loss of 90 percent at each step.

TROPHIC LEVELS AND EFFICIENCY OF FOOD CONVERSION

If more food cannot be produced, consideration can be given to cutting waste. Waste can be reduced by consuming foods as close to the beginning of the food chain as possible. As suggested earlier, vegetable products can be substituted for meats. This might save up to 90 percent of the food generally lost through waste, catabolism, and excretion. Therefore, only about 10 percent of the original vegetable matter is actually converted into the substance of an animal's body.

The extent of food loss in long food chains may be illustrated by the chain beginning with phytoplankton of the ocean as the producing level. Feeding on the plankton is a diverse group of primary consumers including all sizes of animals from zooplankton to whales. When man eats large fishes that eat middle-size fishes that eat little fishes, he may be as many as five steps, or trophic levels, removed from the primary producers. Let another animal replace man in the chain, and the chain may be further lengthened if that animal is eaten by still another one. If a chain consists of several steps and the loss at each step is 90 percent, the energy available for the top consumer is considerably reduced (Table 12–1).

REFERENCES AND FURTHER READINGS

Borgstrom, G. 1967. *The Hungry Planet*. Rev. ed. Collier, New York.

Bresler, J. B., ed. 1966. *Human Ecology*. Addison-Wesley, Reading, Mass.

Brown, L. R., and G. W. Finsterbusch. 1972. *Man and His Environment: Food*. Harper & Row, New York.

Dasmann, R. F. 1964. *African Game Ranching*. Pergamon–Macmillan, New York.

Johnson, C. E., ed. 1970. *Human Biology: Contemporary Readings*. Van Nostrand Reinhold, New York.

Scientific American. 1970. "The Biosphere; Readings from Scientific American." Freeman, San Francisco.

Paddock, W. and P. Paddock. 1967. *Famine—1975! Little, Brown, Boston*.

Chapter 13
Exploitation

NATURAL RESOURCES

The early settlers of the Western Hemisphere and other virgin areas found the lands spacious, rich, and unprotected. Anything they found was theirs for the taking, and they often took it in a wasteful manner. To them the resources must have seemed limitless compared to those they had left behind. Their descendants, a few generations removed, are finding it hard to come to grips with the reality that demand is overtaking supply, as people in overcrowded parts of the world are already well aware.

The fundamental natural resources are air, water, and soil. Without them life would quickly perish. Their value is easily underrated because they are abundant and seemingly inexhaustible; their real value should be based on their importance rather than their quantity. The danger lies not so much in the possibility that they will be exhausted as in the probability that they will become contaminated or abused to the extent that their availability in unpolluted forms will be jeopardized. The quality of all three has suffered from pollution, but fresh water is the most endangered. Soil has also suffered depletion through misuse and erosion.

Other natural resources, needed more for convenience and comfort than for survival, are forests, wildlife, minerals, and fossil fuels. These vary greatly in quantity, value, and distribution, and world reserves are diminishing every day. Of all these resources, only the living ones are renewable. Eventually the others will be exhausted, and man will have to recycle, substitute, or do without.

EXPLOITATION OF NATURAL RESOURCES

Exploitation means use. To many people it means abuse, and this is in keeping with their observations and experiences. To others it signifies depletion, as in the extractive (mining) industries. Not only does mining appear final insofar as resource loss is concerned, but landscape scars are often left as a reminder.

EXPLOITATION OF WATER

Since water has so many uses, demand for it is on the verge of exceeding supply. In recent years consumption of water has increased almost 2 percent annually. It has increased in all three categories of use: domestic, agricultural, and industrial. The only reasonable prospect for meeting future demand is to reduce consumption, find additional sources such as sea water, distribute surplus water to places where it is needed, or purify and reuse it over and over. All of the methods for increasing the supply are expensive.

Since neither water nor population is uniformly distributed, some locations are threatened much more than others by shortages. The situation is most critical in the Southwest, California, and some of the Middle Atlantic states. Many places can expect rationing during a prolonged drought.

When water must be used repeatedly, it can no longer be regarded as a vehicle for transporting waste, at least untreated waste. Nor can it be of such poor quality that it endangers fishes and other aquatic life.

Not all the water used by man is replaced by precipitation. A considerable amount of the water being consumed is fossil water that cannot be replaced in the foreseeable future. This water comes from artesian wells that tap aquifers or from groundwater (Fig. 13–1). As the water of an aquifer is drained away, the pressure at the surface is progressively lowered. In one large area of southern California where groundwater is used, the level has been lowered 200 feet, and in some places near Phoenix, Arizona, the level has been lowered some 15 feet per year. Obviously, that kind of drainage cannot continue without exhausting the supply. This has already happened in some desert places, causing the once verdant croplands irrigated by fossil groundwater to revert to desert. The porous, sedimentary underground strata of arid lands cannot be recharged for years to come. A few relatively small areas where the underground structure is favorable have been recharged with sewage water from cities, a potentially risky practice.

FIGURE 13–1 *Underground sources of water. (a) Groundwater over impervious rock stratum or bedrock. The upper level (water table) fluctuates from year to year, depending on rainfall, drainage, and withdrawal. (b) Artesian well. The water comes from a porous rock stratum (aquifer) located between two impervious strata. Water enters the aquifer at a higher elevation, where the aquifer is exposed at the surface.*

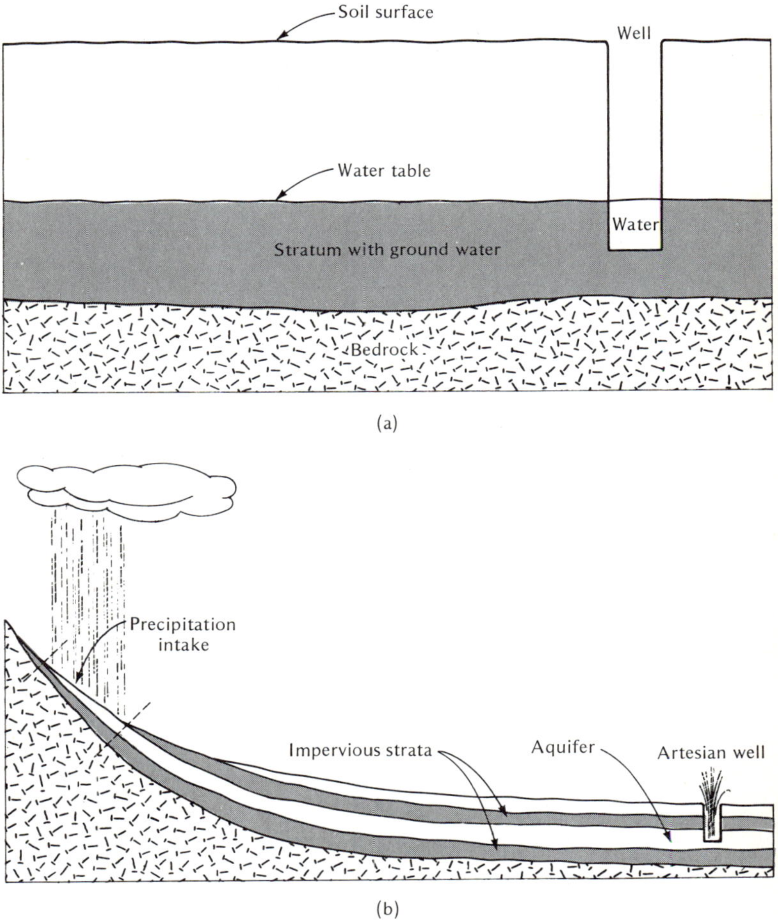

EXPLOITATION OF FORESTS

The overall importance of a forest far exceeds the value of the timber it produces. It stabilizes soil, influences weather, absorbs rainfall, supports wildlife, gives aesthetic satisfaction, and provides recreational opportunities. Its multiple uses are often taken into consideration in forest management practices.

Harvested trees go mainly to lumber, pulp, firewood, charcoal, distillation products, and specialty items (e.g., tool handles or weaving shuttles). Harvesting is still wasteful and destructive owing to the difficulty of handling the product and the large amount of leftovers that cannot be economically salvaged. A falling tree comes crashing down, breaking everything in its path. "Snaking" or hauling logs out of the forest requires the destruction of additional vegetation to clear a pathway. The litter left behind is tinder to fuel a fire as well as an obstruction to those fighting it.

Much wood goes to waste because of the lack of a market or because it is too expensive to transport to market. Wood is no longer widely used for heating or cooking in this country, and diseased trees or trimmings left over from lumbering that could be salvaged for that purpose have few other uses. As in the pioneering days, many private owners of forests who need to clear the land find that the expedient way is still to cut and burn the trees. A tree nurseryman tells of a special kind of waste in which a farmer cut down a stately tulip tree just to harvest its seeds. He sold the seeds to the nursery for a little over $100 and left the wood to rot.

A controversial harvesting practice is clear cutting. Unlike selective cutting, clear cutting harvests all marketable trees at the same time. The area is then replanted, and the trees come to maturity in a pure stand. This procedure keeps down the competition of undesirable trees and facilitates harvesting but leaves the country scarred and ugly after the trees are removed. Moreover, in pure stands, as in other monocultures, there is a tendency for insect parasites and plant pathogens to occur in epidemic proportions. In addition, clear cutting leaves the soil exposed and subject to considerable erosion and removal of nutrients. If the forest is to be put to multiple uses, clear cutting is not nearly as good as selective cutting.

The United States is more than holding its own in maintaining wood reserves. The quality of the reserves, however, may be declining somewhat because of the selective removal of marketable trees while the culls are left standing. In the West new growth is not keeping up with the harvest, whereas in the East and South new growth is exceeding it. The pines of the South grow very fast in the warm, humid climate compared to the firs and spruces of the cold northern coniferous forest.

EXPLOITATION OF WILDLIFE

Wildlife has experienced precarious times because of man's relentless stalking and his destruction of natural habitats. Man has always hunted with little regard to what may happen to the species. Much of the killing is for sport, whose rewards are a story to tell or a trophy to show. Fighting or killing is a way of life among animals, but tooth and claw are no match for a high-power rifle. A number of people cannot understand how a hunter can possibly sacrifice the beauty of a wild beast in the brush for a limp carcass in a pool of blood.

To the sorrow of conservationists, a number of wild animals have been eliminated by man. The majestic herds of buffalo that roamed the plains of the United States are all but gone, preserved like relics in a few protected areas. The only evidence of their former presence is their wallows (Fig. 13–2) and an occasional bone. The passenger pigeon has been completely exterminated. This species was once so abundant that a single flock was estimated to contain over two billion birds and to be over 200 miles long and a mile wide. Since 1700 the extinction of over 130 birds and mammals has been noted, and many others are seriously endangered (Table 13–1).

Most animal populations are endangered only because man is an intruder in the natural scheme of things. Left alone, animal relationships are in the nature of symbiosis in which a kind of homeostasis is maintained. Man upsets this balance by overkilling for food, sport, or profit; by altering the habitat enough to destroy food sources or breeding and shelter sites; or by introducing competitors, diseases, or predators. Wildlife management extends beyond the

FIGURE 13–2 *Buffalo wallow. Shown are the remains of an old wallow used by buffalo when they roamed the prairies of the central United States.*

TABLE 13–1 *Endangered vertebrate species of the United States* [a]

Mammals	
Hawaiian hoary bat	*Lasirus cinereus semotus*
Indiana bat	*Myotis sodalis*
Delmarva Peninsula fox squirrel	*Sciurus niger cinereus*
Morro Bay kangaroo rat	*Dipodomys heermanni morroensis*
Salt marsh harvest mouse	*Reithrodontomys raviventris*
Eastern timber wolf	*Canis lupus lycaon*
Red wolf	*Canis rufus*
San Joaquin kit fox	*Vulpes macrotis mutica*
Black-footed ferret	*Mustela nigripes*
Florida panther	*Felis concolor coryi*
Florida manatee (sea cow)	*Trichechus manatus latirostris*
Key deer	*Odocoileus virginianus clavium*
Columbia white-tailed deer	*Odocoileus virginianus leucurus*
Sonoran pronghorn	*Antilocapra americana sonoriensis*
Birds	
Hawaiian dark-rumped petrel	*Pterodroma phaeopygia sandvicensis*
California least tern	*Sterna albifrons browni*
Hawaiian goose (nene)	*Branta sandvicensis*
Aleutian Canada goose	*Branta canadensis leucopareia*
Laysan duck	*Anas laysanensis*
Hawaiian duck (koloa)	*Anas wyvilliana*
Mexican duck	*Anas diazi*
Brown pelican	*Pelecanus occidentalis*
California condor	*Gymnogyps californianus*
Florida everglade kite (snail kite)	*Rostrhamus sociabilis plumbeus*
Hawaiian hawk (io)	*Buteo solitarius*
Southern bald eagle	*Haliaeetus leucocephalus leucocephalus*
American peregrine falcon	*Falco peregrinus anatum*
Arctic peregrine falcon	*Falco peregrinus tundrius*
Attwater's greater prairie chicken	*Tympanuchus cupido attwateri*
Masked bobwhite	*Colinus virginianus ridgwayi*
Whooping crane	*Grus americana*
Yuma clapper rail	*Rallus longirostris yumanensis*
California clapper rail	*Rallus longirostris obsoletus*
Light-footed clapper rail	*Rallus longirostris levipes*
Hawaiian gallinule	*Gallinula chloropus sandvicensis*
Hawaiian coot	*Fulica americana alai*
Eskimo curlew	*Numenius borealis*
Hawaiian stilt	*Himantopus himantopus knudseni*

TABLE 13–1 (*Cont.*)

Birds (*Cont.*)	
Puerto Rican plain pigeon	*Columba inornata wetmorei*
Puerto Rican parrot	*Amazona vittata*
Ivory-billed woodpecker	*Campephilus principalis*
Red-cockaded woodpecker	*Dendrocopus borealis*
Hawaiian crow (alala)	*Corvus tropicus*
Small Kauai thrush (puaiohi)	*Phaeornis palmeri*
Large Kauai thrush	*Phaeornis obscurus myadestina*
Molokai thrush (Olomau)	*Phaeornis obscurus rutha*
Nilhoa millerbird	*Acrocephalus kingi*
Kauai oo (oo aa)	*Moho braccatus*
Crested honeycreeper (akohekohe)	*Palmeria dolei*
Hawaiian akepa (akepa)	*Loxops coccinea coccinea*
Maui akepa (akepuie)	*Loxops coccinea ochraceu*
Oahu creeper (alauwahio)	*Loxops maculata maculata*
Molokai creeper (kakawahie)	*Loxops maculata flammea*
Akiapolaau	*Hemignathus wilsoni*
Kauai akialoa	*Hemignathus procerus*
Kauai and Maui nukupuus	*Hemignathus lucidus*
Laysan and Nihoa finches	*Psittirostra cantans*
Ou	*Psittirostra psittacea*
Palila	*Psittirostra bailleui*
Maui parrotbill	*Pseudonestor xanthorphrys*
Bachman's warbler	*Vermivora bachmanii*
Kirtland's warbler	*Dendroica kirtlandii*
Dusky seaside sparrow	*Ammospiza nigrescens*
Cape Sable sparrow	*Ammospiza mirabilis*
Reptiles and Amphibians	
American alligator	*Alligator mississippiensis*
Blunt-nosed leopard lizard	*Crotaphytus silus*
San Francisco garter snake	*Thamnophis sirtalis tetrataenia*
Puerto Rican boa	*Epicrates inornatus*
Santa Cruz long-toed salamander	*Ambystoma macrodactylum croceum*
Texas blind salamander	*Typhlomolge rathbuni*
Houston toad	*Bufo houstonensis*
Fishes	
Shortnose sturgeon	*Acipenser brevirostrum*
Longjaw cisco	*Coregonus alpenae*
Lahontan cut-throat trout	*Salmo clarki henshawi*

TABLE 13–1 (*Cont.*)

Fishes (*Cont.*)	
Piute cut-throat trout	*Salmo clarki seleniris*
Greenback cut-throat trout	*Salmo clarki stomias*
Gila trout	*Salmo gilae*
Arizona (Apache) trout	*Salmo sp.*
Humpback chub	*Gila cypha*
Mohave chub	*Siphateles mohavensis*
Pahranagat bonytail	*Gila robusta jordani*
Moapa dace	*Moapa coriacea*
Woundfin	*Plagopherus argentissimus*
Colorado River squawfish	*Ptychocheilus lucius*
Kendall Warm Springs dace	*Rhinichthys osculus thermalis*
Cui-ui	*Chasmistes cujus*
Devil's Hole pupfish	*Cyprinodon diabolis*
Comanche Springs pupfish	*Cyprinodon elegans*
Tecopa pupfish	*Cyprinodon nevadensis calidae*
Warm Springs pupfish	*Cyprinodon nevadensis pectoralis*
Owens River pupfish	*Cyprinodon radiosus*
Pahrump killifish	*Empetrichythys latos*
Big Bend gambusia	*Gambusia gaigei*
Clear Creek gambusia	*Gambusia heterochir*
Pecos gambusia	*Gambusia nobolis*
Unarmored threespine stickleback	*Gasterosterus aculeatus williamsoni*
Gila topminnow	*Poeciliopsis occidentalis*
Fountain darter	*Etheostoma fonticola*
Watercress darter	*Etheostoma nuchale*
Maryland darter	*Etheostoma sellare*
Blue pike	*Stizostedion vitreum glaucum*

[a]Source: U.S. Department of the Interior.

protection of game animals from hunters to the protection of habitats as well.

Hunting is a problem that requires careful regulation. Some people even hunt from automobiles along the highways or from planes. But hunting is not always solely destructive. It can be useful in controlling population size. Whether it is good or bad depends on the density of the population and the ability of the habitat to support it. Population size can be regulated by controlling the size of the daily take or by varying the length of the hunting season.

Changing the number of animals in an ecosystem involves a great deal of risk. Without the natural enemies to which it is accustomed, an introduced species often becomes pestiferous and difficult to control. It may be a serious enemy to indigenous species that are not adapted to defend themselves. Some marsupials of Australia, for example, have fared poorly against introduced felines and canines. Flightless birds that evolved in isolated places where there were no predators have been lucky to survive the introduction of some carnivores.

Like the introduction of a new species, the removal of a predator upsets the balance because it allows a native species to get out of hand. Unregulated by predators, the animals multiply until there is a shortage of food; then they may die from starvation or migrate into new areas. The bounty system, which is intended to protect domesticated herds in some of the Western states, not only accomplishes its objective but allows other animals like rodents to increase in number. In turn, these animals require more food, and this taxes the seed supply of some plants to the extent that they cannot reseed properly. The overall impact of killing the predators is probably negative. Most naturalists agree that predators serve a useful function by making room for more vigorous animals when they kill the old, sickly, or inferior ones.

Considering all of the pressures on wildlife, it is surprising that animals have survived as well as they have. Actually, many of them are impossible to eradicate, and there is a continuous battle against a few troublesome ones. Though they are vigorously opposed, the Japanese beetle and the fire ant continue to extend their ranges. In general, the large animals are the most endangered because they tend to be fewer in number and are easier to hunt.

Greatest attention has been given to saving animals that are hunted—game mammals, birds, and fishes. In this country such efforts have been largely successful. In some sections of the country, animal populations are much higher than they have been in many years. The hunter, moreover, is encountering more difficulty in finding places to hunt because many property owners will not allow trespassing.

Fortunately, wildlife is a renewable resource like the forests. If a species is vigorous, it will reproduce rapidly. Some highly specialized animals that eat a single kind of food or require an unusual habitat are at a distinct disadvantage. A few, like the California condor, may not survive because of an inherent reproductive weakness: The female condor lays only one egg every other year.

A positive attitude toward wildlife preservation seems to be emerging in many quarters. The collective judgment of society is repulsed by wanton killing and is tightening the rules. More people are refusing to buy animal products like fur coats or alligator bags. Such practices as killing elephants for tusks or rhinos for horns have been all but abandoned. On the popular African safaris the camera is rapidly replacing the gun. There and elsewhere, more and more people are discovering the beauty of wildlife. Especially in national parks, people flock around wild animals with interest and curiosity. Even the little chipmunks get their share of attention as children and adults alike feed them and patiently await the thrill of having them take a tidbit directly from the hand.

EXPLOITATION OF LAND

It is well known that the land of the United States was exploited drastically as it was put to the plow after the initial exploration period. So extensive was the clearing of land that large tracts of the original vegetation were virtually destroyed. The erosion that followed was especially acute on arid lands, which periodically experience severe drought, and on steep slopes unsuited for cultivation.

The tragedy of erosion is that the topsoils, which take a long time to develop, are the first to go. In eastern America, where rainfall is relatively high, the time required to develop an inch of topsoil is at least 100 years. Therefore, if the soil is eight inches deep, eight centuries are involved in its making. On many fields all the topsoil has been depleted in a single generation.

To conserve soil all land should be put to proper use. Clean-tilled crops should be grown on level land. Cover crops and pasturage are suitable for slopes, rocky soils, or other sites unfit for cultivation. Major faults of the past have been use of land for wrong purposes, failure to practice contour farming and terracing, and destruction of the ground cover that protects and binds the soil. Many farmers have been guilty of clearing land that is unsuitable for the use intended. As soon as unfavorable weather comes, the field becomes a wasteland, either washing away as mud or blowing away as dust. Just as misused farmland is devastated, so is land that is overgrazed.

Of all the land in the fifty states, about 58 percent is crop and grazing land; 22 percent ungrazed forest land; 12 percent "waste" land (deserts, swamps, tundra, etc.); 5 percent parks, wildlife refuges, and other public properties; and 3 percent urban centers, roads, and airports (Fig. 13–3). Rural land is being converted to nonagricultural uses at a rate of about two million acres per year. The converted land is used for wilderness areas, parks, recreation areas, wildlife refuges, urban development, highways, airports, and reservoirs.

In an effort to avoid some of the mistakes of the past, many private and governmental agencies are energetically involved in land-

FIGURE 13–3 *Land use and ownership in the United States. (Source: U.S. Department of Agriculture.)*

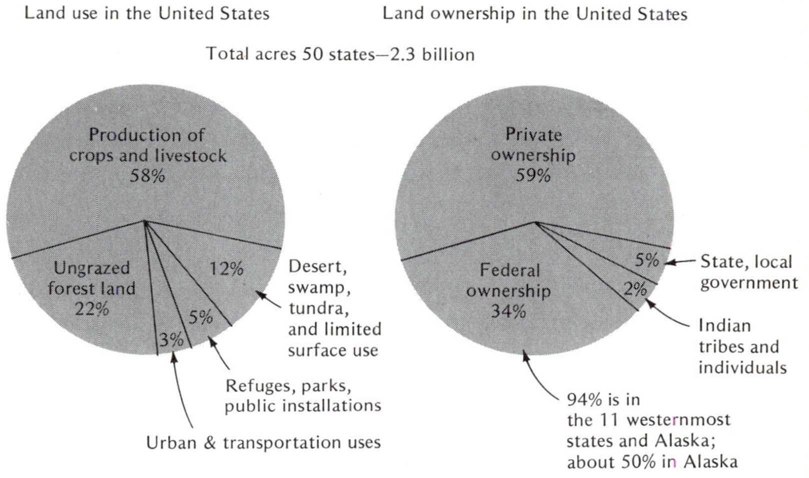

use planning. They have the responsibility of charting the development of properties held by many separate landholders; much of their work is educational. They also have to coordinate the efforts of all participating parties, a job that is complicated by the conflicting interests of individuals, the needs of neighborhood units, and national objectives.

EXPLOITATION OF NONRENEWABLE RESOURCES

Minerals and fossil fuels are nonrenewable natural resources. First to be taken are the high-grade, easily accessible resources, followed by the low-grade or hard-to-get ones that are extracted less economically. The cost of extraction increases constantly until a point of no return is reached. Some known reserves are already on the verge of exhaustion. According to the geologist Preston Cloud, at the present rate of consumption the United States will have exhausted its known high-grade reserves of oil, natural gas, uranium, manganese, chromium, nickel, tungsten, cobalt, copper, lead, zinc, aluminum, gold, and silver by the turn of the century. Of the three major ingredients in fertilizers, phosphorus is the one that is in shortest supply.

The depletion problem is compounded by the growing demands of an ever-increasing population. Energy requirements are already high and increasing daily. Gas and oil seem sure to run out first.

After the wells go dry, oil can be extracted from oil-bearing shales, or coal can be converted to a liquid fuel. These valuable resources, though abundant, are far from limitless.

The unequal distribution of the earth's resources leaves all nations deficient in some of them. Even the United States depends on other countries for many raw materials. For example, it is a heavy importer of such common minerals as iron, copper, and nickel. If other countries consumed as much of these materials on a per capita basis as Americans, the supply would be used up instantly.

Future extraction of earth resources will require many changes in methods. Mines will go deeper into the crust, and the ocean floor will be tapped as never before. New mining techniques will be developed for easier extraction and handling of resources. With these changes must come the environmental rehabilitation that has so long been neglected.

Two great problems of rehabilitation concern strip-mine operations and tailings. Strip mining and open-pit mining (Fig. 13–4) have the big advantage of being more economical than underground mining. These operations are feasible where the overburden (overlying strata) is thin. Because earth moving is expensive, the overburden is usually dumped in the most convenient place. After the operation is finished, the countryside is scraped, gouged, and

FIGURE 13–4 *An open-pit copper mine at Aho, Arizona.*

heaped. The original vegetation is gone, and the drainage system has been ruined by erosion. Most states now have laws that require mine operators to refill, contour, and stabilize the land with vegetation. Unfortunately, such laws cannot be applied retroactively to correct the immense damage already done. Even a present-day operator may go bankrupt and leave a job unfinished. The nation and its states have so many problems that they cannot afford to cover the scars of the past—so the wounds persist.

After the desired products are removed from the ore, mine tailings pile up, polluting the air with dust and the streams with acids, other poisons, and sediments. Sometimes noxious substances leach into nearby streams and cause considerable destruction (see Chapter 14). Some are new compounds produced by chemical reactions within the tailing heaps. In other cases, like the radioactive leftovers from uranium mines, they are unaltered.

All mine wastes are unsightly and sometimes dangerously unstable. In 1966 a large slag pile cascaded down a mountain slope at Aberfan, Wales, where the waste from coal processing had been accumulating for almost a century. In this accident 144 people were killed. Another disaster almost as bad occurred on February 26, 1972, along Buffalo Creek in West Virginia. A dam of slag and coal leavings from fifteen years of mining operations collapsed and released a flood of waste water that destroyed or heavily damaged fourteen mining communities in the narrow valley. One hundred and seventeen people were known to be killed, a number were never accounted for, and 3000 to 4000 people were left homeless.

REFERENCES AND FURTHER READINGS

Bureau of Sport Fisheries and Wildlife. 1966. *Rare and Endangered Fish and Wildlife of the U.S.* Resource Publ. 34, Supt. Doc., Washington, D.C.

Detwyler, T. R., ed. 1971. *Man's Impact on Environment.* McGraw-Hill, New York.

Fisher, J., N. Simon, and J. Vincent. 1969. *Wildlife in Danger.* Viking, New York.

Greenway, J. C., Jr. 1967. *Extinct and Vanishing Birds of the World.* 2d rev. ed. Dover, New York.

Wagner, R. H. 1971. *Environment and Man.* Norton, New York.

Chapter 14
Pollution of Air and Water

POLLUTION, AN ABNORMAL ENVIRONMENTAL CONDITION

Whenever the natural environment is altered by impurities or artificial conditions there is the risk that some organisms will suffer and the possibility that others will benefit. This being the case, particular care should be used in defining the term **pollution**. One viewpoint is to call pollution any abnormal environmental condition that upsets a natural equilibrium. Actually, most people expand the meaning to include anything they dislike, such as bad odors or unsightly litter. When the environmental alteration is more beneficial than harmful, at least to the person receiving the benefits, it is sometimes called **enrichment** instead of pollution.

No environment is forever clean because nature itself is both a purifier and a polluter, depending on time and circumstances. Many natural events like volcanic explosions or dust storms add particulate matter to the atmosphere. Even plants and animals pollute. At least from man's point of view, the stinking seeds of a ginkgo tree are pollution. An example of pollution from the animal world is the excrement of birds. Where they congregate to roost, their droppings often kill the vegetation. The way pigeons and other birds pollute urban areas is too familiar to need amplification. And now man stands self-condemned for having become the greatest polluter of all.

Since contamination of some sort is an ever-present possibility, its detection and measurement are important as long as mankind has within his power the means to control it. He needs to know the consequences of environmental impurities at different levels of concentration and to determine the point at which the situation becomes intolerable. In the language of the ecologist, he must know the **tolerance level** for each contaminant, the so-called range of safety where damage is nonexistent or minimal. He must then see that the tolerance level is not exceeded.

The heart of the problem is man's tendency to waste. He has almost always taken what he wanted at the moment and discarded in the easiest way what he did not want. He has discarded waste individually and collectively to the extent that everybody is guilty in some manner. There is no basic difference between an individual who throws trash in the street and a large company that dumps chemicals in a stream; both are selfish and inconsiderate.

Waste should be considered a resource and utilized as much as possible. Some materials can be used just as they are or can be reprocessed; many can be converted to other uses such as mulch, fertilizers, fuels, or building materials. Because pollution is beginning to endanger life, the most expedient waste disposal method may be the most expensive in the long run.

AIR POLLUTION

Air pollution is most acute near places where pollutants are generated. Generally this means inside factories and mines or around large cities and industrial complexes. Although the people who live and work in such places bear the brunt of the damage, everybody is affected in that the air and its contaminants circulate the globe.

The phenomenon of temperature inversion sometimes traps a layer of heavily polluted air at the surface of the earth for long periods, preventing its dissipation elsewhere. In an inversion a layer of cold air, sometimes polluted, underlies a warm layer (Fig. 14–1), reversing the normal situation. When the cold layer of polluted air is trapped in a basin, the condition may take a long time to correct itself. In Los Angeles between June and October, the layers of air are inverted 80 to 90 percent of the time.

The problem of inversion and undissipated smog is primarily a matter of health, not merely discomfort. The lungs of city dwellers lose their natural color and become completely blackened with soot. Irritation of the eyes is frequent, but the magnitude of the problem is fully realized when people are forced to remain indoors during an inversion. In some cases deaths are attributed to smog.

Coal miners, textile workers, chemical workers, and cigarette smokers are susceptible to anthracosis (black lung disease), byssinosis (brown lung disease), silicosis (miner's asthma), asbestosis,

FIGURE 14–1 *Inversion of layers of air. (a) The normal condition with warmer air below colder air. (b) An inverted condition in which a cold layer, in this case polluted, is trapped below a warm layer.*

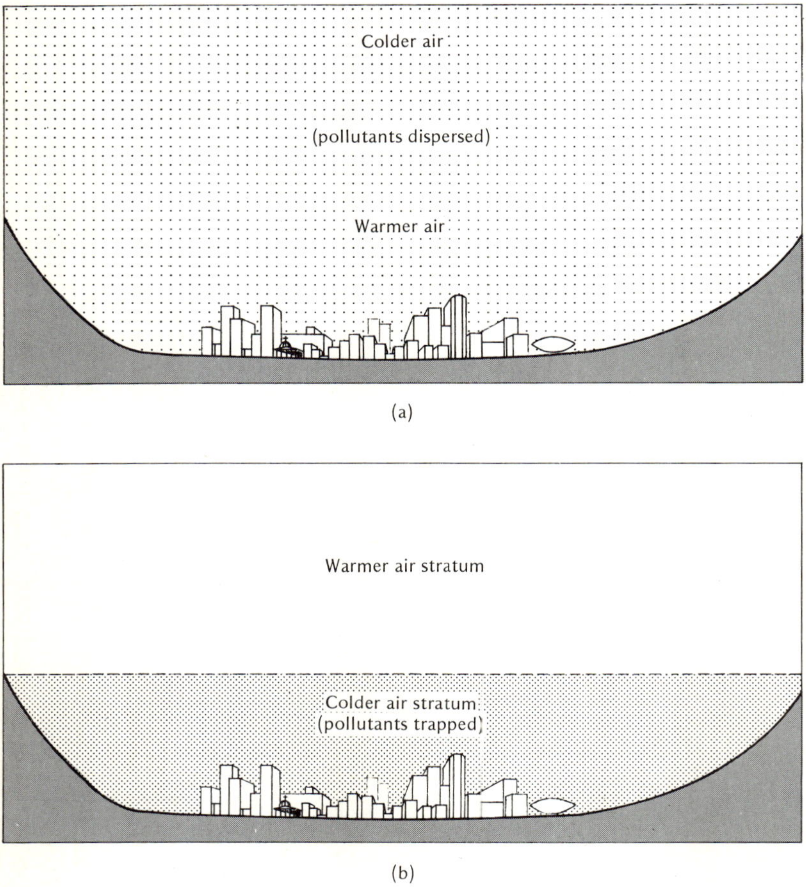

emphysema, and cancer as a result of air pollution. Unfortunately, such long-term damage is almost entirely irreversible. Damage to the ecosphere is generally temporary but may require a considerable period of time for recovery.

The Copper Basin of southeastern Tennessee is a spectacular example of atmospheric pollution and the length of time required for recovery from the damage it causes (Fig. 14–2). The devastation of the area started shortly after the discovery of copper in 1843, and by 1880 about fifty square miles of mountain land lay desolate, devoid of vegetation and badly eroded. The early companies stripped the timber from the land and used it in the copper-extracting process. They alternately stacked timber and ore in a large pile and set it afire to smelt the metal. Sulfur dioxide from the fires killed what vegetation was left. The land, mountainous and subject to heavy rainfall, was eroded beyond description.

Extensive and persistent efforts have been made to repair the damage. Since 1907 the Tennessee Copper Company (now a subsidiary of Cities Service Corporation) has been converting sulfur dioxide, which formerly did so much damage, to sulfuric acid, now the company's main product. No longer subjected to poisons from the air, the vegetation has had a chance to recover. However, the soil lost through erosion is gone forever.

Efforts to reclaim the land were initiated by the Civilian Conservation Corps in the mid-1930s. To date, more than seven million trees—loblolly pines (*Pinus taeda*)—have been planted; many have died. At first the Tennessee Valley Authority (TVA) supplied the tree seedlings, but since 1954 the copper company has assumed that responsibility. The plan is to concentrate reforestation efforts at the periphery of the blighted area and work inward. Grasses, a Japanese knotweed (buckwheat), and kudzu have been planted around the

FIGURE 14–2 *Copper basin. A scene from the Copper Basin near Ducktown, Tennessee.*

company grounds. The knotweed is particularly good for plugging gullies. Other plants have been tried, but without success.

The basin still looks ruined (Fig. 14–3), but old residents say they can see progress. The company continues its efforts at reclamation, with the objective of planting half a million tree seedlings annually.

There are many kinds of air pollutants. Most of them do not leave such spectacular scars as the Copper Basin. Some, however, may well be more destructive in their total impact. The most important classes of air pollutants are:

1. Particulate matter
2. Carbon monoxide
3. Carbon dioxide
4. Hydrocarbons
5. Nitrogen oxides
6. Sulfur oxides
7. Photochemical oxidants

Particulate matter consists of small particles of all sorts of substances suspended in the air. They may be tiny solid or liquid particles known as **aerosols** or larger particles like dust and droplets. Naturally, some substances are more harmful than others. Collectively, they irritate animals, poison life, cause bad odors, clog the breathing pores of plants, block visibility, mar the surfaces of buildings, damage clothes, affect the weather, and perhaps change the

FIGURE 14–3 *A few struggling pine trees in the Copper Basin near Ducktown, Tennessee. Considerable efforts have been made to revegetate the area.*

climate. Such substances were polluting the air in ever-greater amounts in this country until recently, when federal and state controls were tightened.

Lead is a toxic particulate substance produced primarily by the automobile, which burns gasoline with lead additives. Symptoms of chronic lead poisoning are loss of appetite, weakness, listlessness, awkwardness, and sometimes convulsions. Many people have accumulated lead in quantities just short of that needed to produce symptoms of poisoning. Those who have had the greatest exposure to automotive exhaust, such as traffic policemen, have larger concentrations of lead in their blood than the average man. Nobody is protected, however; lead has been found in polar ice far removed from where it was produced. An estimated 200,000 tons of lead is discharged annually in the United States alone.

Particles in the air block sunlight from reaching the earth's surface and thus may actually cool the earth. Their effectiveness in excluding light is evident in the smog blankets that occur over cities. Smog (word coined from smoke and fog) is a mixture of visible particulate matter and associated but invisible waste. The basic particulate matter in smog comes from the burning of leaves, wood, sawdust, shavings, paper products, coal, oil, and gas. It consists largely of soot (the carbon particles of smoke) and fly ash, which consists of noncombustible particles derived from burning. Dust, unlike the carbon and fly ash particles of smog, is a natural pollutant derived from the cultivation of soil, dust storms, and volcanic explosions. In addition, several kinds of dust are produced by industrial operations.

In 1883 the violent volcanic explosion of Krakatoa threw so much dust into the atmosphere that more than a year was required for natural processes to remove it. There was so much dust in the air immediately after the explosion that lamps had to be used during the day in Batavia, 100 miles away. World temperatures were noticeably lowered by the dust blanket. If the estimate that atmospheric turbidity has increased tenfold in recent years is correct, a cooling effect is inevitable unless other conditions counteract it.

Besides cooling the earth, particulate matter seeds clouds by supplying nuclei for the condensation of water. La Porte, Indiana, illustrates this phenomenon. Located about thirty miles downwind from large industrial centers, since 1865 it has received 30 percent more rain than adjoining communities not directly in the path of the particulate cloud. Although the consequences of a seeded cloud are locally recognized, the degree to which it reorders the natural patterns of rainfall distribution is not clearly discernible.

Smog, which contains other substances in addition to particulate matter, differs from place to place in composition and concentration. At its worst it can kill, as in London during a four-day period in December, 1952, when about 4000 people died because of it. The damage often extends beyond the big cities into the countryside. According to the California State Department of Agriculture, air pollutants cost the citrus growers of the Los Angeles basin $25.7 million in 1970 and $44.5 million in 1969. Los Angeles smog is also killing some types of desert vegetation in the Coachella Valley and ponderosa pines in the San Gabriel Mountains. Even as far away as the San Jacinto Mountains, some 100 miles from Los Angeles, ponderosa pines are damaged up to elevations of 7500 feet.

Carbon monoxide (CO) is a gaseous product of incomplete combustion coming chiefly from gasoline engines. An estimated two-thirds of the carbon monoxide released in the United States comes from the automobile (Table 14–1). The amount of carbon monoxide in a metropolitan area can be directly correlated with the amount

TABLE 14–1 *Carbon monoxide emission estimates by source—1968*[a]

Source	CO emissions (10^6 tons/yr)	Percentage of total
transportation	63.8	62.8
motor vehicles	59.2	58.2
gasoline	59.0	58.0
diesel	0.2	0.2
aircraft	2.4	2.4
vessels	0.3	0.3
railroads	0.1	0.1
other nonhighway use of motor fuels	1.8	1.8
fuel combination in stationary sources	1.9	1.9
coal	0.8	0.8
fuel oil	0.1	0.1
natural gas	N	N
wood	1.0	1.0
industrial processes	11.2	11.0
solid-waste disposal	7.8	7.7
miscellaneous	16.9	16.6
man-made	9.7	9.5
forest fires	7.2	7.1
total	101.6	100.0

[a]Source: U.S. Department of Health, Education, and Welfare.

FIGURE 14–4 *Carbon monoxide concentration in an urban area. Shown is the relationship of carbon monoxide concentration to traffic levels in midtown Manhattan. (Source: U.S. Department of Health, Education, and Welfare.)*

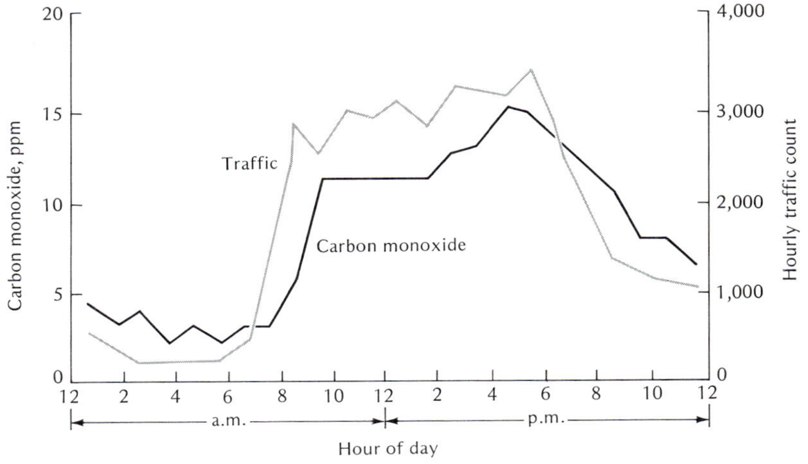

of traffic (Fig. 14–4). In New York City it is about equal in quantity to particulate matter, sulfur dioxide, nitrogen oxides, and hydrocarbons combined, or about 1.5 million tons per year.

Carbon monoxide reduces the oxygen supply of humans by combining with some of the hemoglobin that should be absorbing and transporting oxygen. Early symptoms of carbon monoxide poisoning are lack of energy, dizziness, headache, nausea, decreased sensitivity, and lack of coordination. More extreme symptoms are convulsions and loss of consciousness. If the dosage is large enough, death by asphyxiation follows. Smoking in confined places produces enough carbon monoxide to cause some of the early symptoms of poisoning. The carbon monoxide absorbed from smoking and living in a heavily contaminated environment raises the level of the gas in the body to dangerous proportions. People have died because of carbon monoxide emitted from malfunctioning heaters and furnaces.

Carbon dioxide (CO_2) is a natural constituent of the atmosphere and as such is not a pollutant except in special situations. It is part of the carbon cycle and is fixed into foods by green plants. The normal atmospheric content is about 0.03 percent, and it can be increased to as much as thirty times that amount before the photosynthetic rate is at its maximum. Two major processes, respiration and combustion, release carbon dioxide into the air. It has been calculated that one person releases about 23 liters of carbon dioxide per day through respiration. With the coming of the industrial age and the burning of fossil fuels, carbon formerly locked in these substances has been released as carbon dioxide.

In addition to temporary and local situations in which too much carbon dioxide is harmful to life, the gas can potentially change the climate. Some scientists are concerned that increasing quantities of carbon dioxide may accumulate in the troposphere, where it would trap heat near the surface of the earth the way glass traps heat in a greenhouse. Their reasoning is a follows: Incoming shortwave solar radiation passes through the atmosphere essentially unimpeded and heats anything that absorbs it. Thus, shortwave energy is converted into heat energy, or longwave (infrared) energy. The reradiated long waves are known to be absorbed and reflected by water vapor in the atmosphere, as they are by the glass of a greenhouse. If carbon dioxide also tends to block the reradiation of long wavelengths, one would expect the atmosphere to become warmer as the concentration of atmospheric carbon dioxide increases.

Atmospheric carbon dioxide has increased a little over 13 percent in the past hundred years. Other studies show that from 1958 to 1970 the carbon dioxide concentration increased 0.7 part per million per annum. If the same rate of increase continues, atmospheric carbon dioxide will double in approximately twenty-three years. From 1965 to 1969 only about 40 percent of man-made carbon dioxide stayed in the atmosphere. The remainder must have been absorbed by the ocean or fixed into foods by green plants.

Should the present rate of increase in carbon dioxide continue, the atmosphere will retain a higher proportion of heat than ever before. This will melt polar ice caps and raise the level of the oceans. Naturally, it will also affect the kinds and distribution of life on the earth. Over a fifty-year period from 1890 until 1940, the mean temperature increased 1.1°F. Since 1940, however, the temperature has dropped about 0.5°F despite the rise in atmospheric carbon dioxide. The counteracting trend is probably due to the increase in particulate matter that can block incoming radiation.

Hydrocarbons (HC) are unburned fuels. Over half of the hydrocarbons causing air pollution are wasted by vehicles used in transportation. More than 125 hydrocarbons have been specifically identified. They come not only from engine waste but also from the evaporation of fuels in tanks and carburetors. Since hydrocarbons themselves are not harmful at concentrations normally found in the atmosphere, their greatest danger is in the products derived from them by photochemical activity.

Nitrogen oxides (NO_x) are produced by fuels burning at high temperatures, as in large boilers and internal combustion engines.

High temperatures cause the nitrogen to combine with oxygen. The amount can be reduced considerably if exhaust gases containing it are not allowed to cool quickly. There is some evidence that these oxides may be harmful to human health. Like the hydrocarbons, nitrogen oxides produce a second generation of chemicals through photochemical activity. Many of these secondary products are definitely harmful.

Sulfur oxides (SO_x) are released by burning fuels that contain sulfur. They are extremely poisonous to plants, as illustrated by the destruction of nearly all vegetation in the Copper Basin (Fig. 14–3). Animals are also seriously affected; they suffer most from injury to the delicate tissues of their respiratory systems. Inanimate objects are damaged by corrosive action. At present, about 60 percent of sulfur oxide contamination comes from the combustion of oil fuels and from industrial processes using sulfur.

Photochemical oxidants are derivatives of other contaminants like nitrogen oxides and hydrocarbons. The energy from sunlight activates chemical reactions that produce many secondary pollutants. Among these are ozone, aldehydes, organic acids, peroxyacyl nitrates, sulfuric acid, and peroxides. The quantities of these oxidants in the air vary with location, time of day (Fig. 14–5), and season. Smog is a veritable factory of photochemical activity.

These photochemical air pollutants are known to produce measurable effects on both living and nonliving matter. For example, the relationship of concentration and effect of ozone is given in Table 14–2. Injury to vegetation is one of the earliest manifestations of such pollution. In fact, the sensitivity of certain plants to these substances makes them very useful biological indicators of photochemical air pollution. Pinto beans and petunias are used to monitor the presence of peroxyacyl nitrates, while tobacco and annual bluegrass plants are used to show the presence of ozone.

WATER POLLUTION

The physical characteristics of water that permit it to hold many substances in suspension or solution account for the heavy load of undesirable substances it often contains. In fact, it is known as the universal solvent because it dissolves more materials than any other liquid. These characteristics together with its fluidity, which sends it coursing to the sea in streams, make water convenient for flushing waste. However, useful as streams may be for disposal, they are even more important as suppliers of water for drinking, washing, cooling, and irrigation. To be used for these purposes, the water must remain relatively pure.

Pollution has increased to such an extent that many aquatic habi-

FIGURE 14–5 *Photochemical oxidants. Shown is the relationship of photochemical-oxidant production to time of day for a typical summer day in an eastern city. (Source: U.S. Department of Health, Education, and Welfare.)*

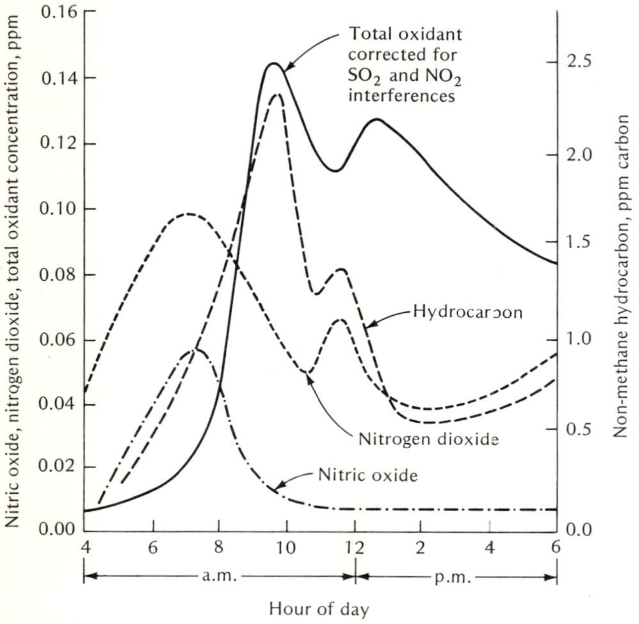

tats are no longer habitable by fishes and other organisms. Some streams are so bad that they are virtually worthless except for drainage and waste removal. Pollution of water originates from many sources, such as sewage, industrial wastes, sediments, fertilizers, manures, oils, and poisons (biocides).

Sewage disposal is primarily a matter of sanitation. Still, much of the world population suffers unnecessarily from odors, vermin attracted to filth, and disease because the problem is ignored or improperly handled. In this country about one-third of the population is not even served by a sewer system, and less than one-third by an adequate one. In general, the problem is worse where people are concentrated. Surprisingly, about 5 percent of the population is served by sewers that discharge raw sewage directly into streams.

Municipal sewerage systems are by no means intended to handle only human waste. They also carry a great deal of kitchen refuse, much of it simply washed down the drain unaltered or pulverized

TABLE 14–2 *Effects of ozone on organisms*[a]

Species	Concentration, ppm	Time	Effects
vascular plants			
peanut			
(*Arachis hypogaea*)	0.02	24–28 hrs	chlorosis; leaves may abscise
radish			
(*Raphanus sativus*)	0.05	8 hrs/day (20 days)	50% reduction in yield
carnation			
(Dianthus *caryophllus*)	0.07	60 days	50% reduction in floral development
tobacco			
(*Nicotiana tabacum*)	0.10	5.5 hrs	50% reduction in pollen germination and pollen tube growth
	0.05	8 hrs/day (20 days)	reduction in growth
bacteria			
Streptococcus salivarious	0.025	30 minutes	90% mortality
Escherichia	4.0	—	retarded growth
coli	10.0	—	prevented growth
Staphylococcus pyogenes and S. aureus	3.0–4.6	1.5–4.5 hrs	no positive germicidal action
protozoa			
Colpidium	8.0	1 hr	100% mortality
Blepharisma			
Paramecium			
Amoeba			
Didinium			
Tillina			

[a]Source: Environmental Health Service, U.S. Department of Health, Education, and Welfare.

by garbage disposals. Added to these are assorted street deposits channeled into the system by storm sewers. Many sewerage systems, especially in larger cities, also handle industrial waste. In fact, almost half the waste processed by treatment plants comes from this source. If these materials are toxic to the bacteria on which treatment depends, they must be diluted.

Raw animal excrement has long been used for fertilizer despite some risks involved. Even now, millions of Asiatic farmers fertilize their crops with human excrement, known as "night soil." Farmers the world over must dispose of barnyard and stable manure, which they deposit on fields to enrich the soil. Much of the manures from farm animals wash from fields, pastures, and feedlots into streams and contaminate them. Someone has estimated that body wastes from domesticated animals in the United States are equivalent in quality to that produced by a population of two billion people. This, of course, goes unprocessed.

Processing sewage is a matter of fast recycling of organic materials. It consists primarily of bacterial decay, often accelerated by providing additional oxygen. Usually, there are one to three distinct steps. The first step, called **primary treatment**, is a gravity process: The sewage is held for a while in tanks so that the solids can settle from the liquids (Fig. 14–6). Simultaneously, bacterial action reduces the organic content by about 25–30 percent. The second step, **secondary treatment**, accelerates decomposition by adding oxygen, either by activating (stirring) the sludge (Fig. 14–7) or by spraying it over a bed of rock covered with bacteria. This process may remove all but 10 to 15 percent of the organic matter. The third step, **tertiary treatment**, consists of one or more of several available processes such as flocculation, filtration, and carbon adsorption. The third system may remove all but 1 percent of the organic residue. What remains of the sewage, mostly water, is discharged in natural bodies of water.

Cost is an important consideration in any sewage disposal sys-

FIGURE 14–6 *Schematic of a sewage treatment facility.*

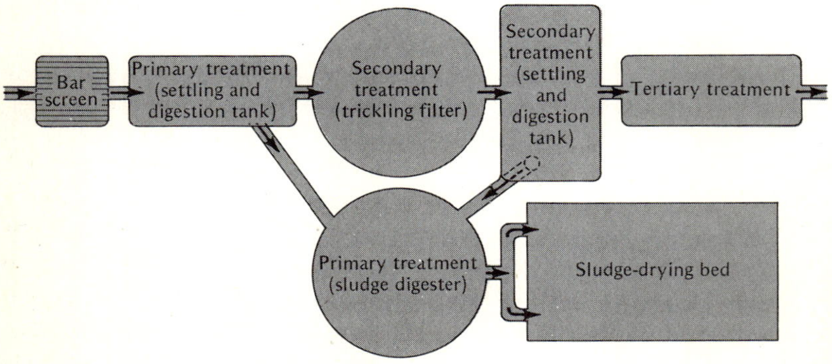

FIGURE 14–7 *A secondary-treatment facility for a town of about 2000 inhabitants. The large pool has two sludge activators to increase the oxygen supply. The small section walled off, called the polishing pool, is a holding pool where reduction of organic matter continues.*

tem. The extent to which the water is purified depends on local conditions and state laws. If the effluent can be dumped into a large stream or lake, dilution quickly diminishes the percentage of waste to volume of water. A small town with little or no industrial waste may have a system that can remove 90 to 95 percent of the degradable content in the primary and secondary steps. The sewerage system of a large city, receiving a high portion of industrial waste, may remove only 85 to 90 percent of the degradable material in the first two steps.

Until recently many cities have not processed sewage beyond the secondary treatment stage. Now that requirements are more stringent, some are adding tertiary treatment. In general, the cost of the tertiary process that increases the purity of water from the 85–90 percent range to the 98–99 percent range is double or triple that of the earlier steps. The removal of the final 1 or 2 percent of contaminants is extremely expensive.

In sparsely populated regions sewage is often "digested" in open lagoons or septic tanks. The lagoons are ponds that should be inoffensive if they are of the proper size for the amount of sewage handled. Septic tanks are underground holding tanks where bacterial action breaks down most of the solids to gases and water. The water drains away through pipes into an absorption field. There the pipes are in short lengths spaced at the joints in order to allow easy seepage into the porous material, such as gravel, in which the pipes are embedded.

Not to be overlooked is the large volume of untreated sewage and other refuse dumped by ships and boats. Until recently few have bothered to complain about it. Some large ships have a complement of operational personnel as large as the entire population of a small village, and the waste they generate is considerable. Large numbers of smaller commercial vessels ply the waters along coastlines and inland waterways, and millions of recreational boats are scattered throughout the country. The sewage, trash, ballast water, and fuel leaks and spills from such craft are often major sources of pollution.

Sewage products (including detergents) and fertilizers contain phosphates and nitrates that are major causes of eutrophication. Although eutrophication stimulates plant growth (Fig. 14–8), which in turn produces oxygen through photosynthesis, the oxygen liberated is often insufficient to support all the plant and animal life

FIGURE 14–8 *Blue–green algae in a stream contaminated with raw sewage.*

in the community. The scarcity is due in part to the large amount of oxygen consumed in the decomposition of organic waste, which is often abundant. Among the first signs of oxygen deficiency is a die-off of fishes. Upon the exhaustion of free oxygen, decomposition is continued by anaerobic bacteria. Instead of releasing carbon dioxide as aerobic bacteria do, they release methane and hydrogen sulfide gases. This is why an extremely polluted stream is discolored and smelly.

Agricultural wastes that pollute streams are of several types. To the sewage products already mentioned should be added fertilizers, poisons, salt, and sediments. Agricultural poisons are **biocides** (herbicides, fungicides, and insecticides). Some are especially dangerous because they do not decompose readily or because they accumulate in organisms where they have a deleterious effect on metabolism. DDT, a chlorinated hydrocarbon, is such a poison. It is found everywhere—in air, water, ice, and soil. And DDT decomposes so slowly that most of what has already been used is still in the ecosphere and will remain there for a long time to come, even after it is no longer used.

One observed effect of DDT is interference with the secretion of eggshells. The eggs thus affected are too fragile to survive the incubation period, thereby threatening the survival of the species. Birds like eagles, brown pelicans, peregrine falcons, and ospreys at the end of the food chain suffer the most because DDT and certain other compounds are increasingly concentrated with each successive step in the chain. The birds get DDT from fishes that have picked it up from organisms at lower levels in the food chain. The accumulation of DDT in a predator at the end of a food chain is like the growth of a savings account—the principal keeps climbing with each additional deposit. DDT accumulates because it is very slowly degraded biologically or chemically and is soluble in the fats of the body. It should be noted, however, that because animals excrete DDT at different rates, some accumulate less than others.

Damage to birds is only one of many known or suspected dangers of DDT. It is feared that this insecticide might be a dangerous poison to humans. It appears, however, that after DDT reaches a certain level in humans additional quantities are excreted. Nevertheless, the amount of DDT held by the body may still be harmful.

Agricultural chemicals, including insecticides, are undeniably valuable in achieving high productivity. Sometimes their value may far outweigh their disadvantages. In any event, their effects should be well known so that a proper judgment can be made about their use. DDT and many other poisons will probably be removed from the market, but others will have to be discovered to replace them.

One should remember that DDT has saved thousands from malaria by controlling its vector, the mosquito.

Erosional products like salts and sediments may produce serious results where they are deposited. Salts are a special problem in arid lands, where they are present in high concentrations in the soil. As mentioned earlier, water drained from irrigated fields may carry enough salts in solution to kill plants growing elsewhere. Silting does obvious damage by filling valuable reservoirs and clogging channels. The silted environments change so much that they are unfit for the biological communities living there. The public has recently become aware of the dangers of destroying vast coastal feeding and nursery marshlands by both natural silting and intentional filling or draining.

The industries that discharge the greatest amount of oxygen-demanding waste (waste degradable by aerobic bacteria) into water are those that produce or refine paper, petroleum, organic chemicals, and metals. Industrial wastes probably amount to three or four times that produced by the portion of the population serviced by sewerage systems. The greatest problem is that these wastes are often so toxic that organisms of decay cannot live in the effluent. According to a U.S. government report, however, most industries can treat waste at a cost of less than 1 percent of gross sales. Some industrial sources claim, however, that the cost is considerably higher.

Among the many industrial pollutants is a class of chemicals called PCB (polychlorinated biphenyls), second only to DDT among synthetic pollutants. They are found in many everyday products like plastics, paints, and rubbers. Like DDT, they have been distributed throughout the world and have accumulated in animals. Highly concentrated in marine fishes, PCB is passed on to anything that uses those fishes for food.

Although it is not ordinarily a pollutant, accidental spillage of oil is an increasing menace to natural habitats. A film of oil on water prevents oxygen absorption and on animals interferes with normal surface functions. Spills from ships or leaks from offshore wells devastate beaches for long periods. Besides killing wildlife and making beaches unfit for recreational purposes, the accidents require expensive cleanup activity.

Mine drainage is another important source of water pollution. Certain minerals in mines and refuse piles react with water and air to form harmful substances. Sulfur compounds from coal and other mines produce sulfuric acid in this way. A great deal of this kind of pollution is found in streams in the coal-mining parts of Appalachia and elsewhere in America. In addition, toxic metals like copper, zinc, and lead are washed from mines.

Aquatic environments may also be damaged by physical changes such as **thermal pollution**. A number of industrial operations use water as a coolant; after picking up heat it is usually returned directly to the most convenient natural body of water (Fig. 14–9). The electric power industry is currently responsible for over 80 percent of the heat pollution in the United States, and the demand for electricity is growing so fast that streams will soon be overburdened if something is not done about it.

Of most recent concern is the increasing number of nuclear generators used by power companies (Fig. 14–10). Such generators produce 50 percent more waste heat per unit than do conventional fossil-fueled plants. They require a large volume of water, which when discharged is 4° to 18°F hotter than its natural temperature. Some power companies have been forced at great expense to build cooling towers to dissipate heat from the water into the atmosphere instead of into the aquatic environment.

The problem, of course, centers around the nature of the impact of changing temperatures on organisms. The water temperature may change upward or downward, but usually it goes up. Where cold-water trout are the native fishes, higher temperatures may exceed their range of tolerance and completely eliminate them. An unusual case in which cooler water changes a stream is illustrated by the discharge of water from Fontana Lake adjoining the Great Smoky Mountain National Park of North Carolina and Tennessee. The cold

FIGURE 14–9 *A steaming sluice from a nuclear reactor. The water flows into a cooling pond before discharge into a natural waterway.*

FIGURE 14–10 *Nuclear generating facility at Keowee-Toxaway near Pickens, South Carolina. Three reactors are under construction at this site, and four additional ones are to be constructed nearby as part of the same complex.*

bottom-water discharged downstream from the lake excludes for miles the warm-water fishes that would normally inhabit the area.

Changes in the aquatic community are by no means limited to the fish populations. An altered environment, in contrast to a stable one, has fewer species but larger populations. The change kills many organisms that cannot tolerate the new conditions, whereas some of the remaining organisms find conditions more favorable. This is exactly what happens in chemical eutrophication, and it would not be amiss to consider thermal pollution a form of eutrophication.

A problem associated with thermal pollution is that of keeping the temperature constant for a stable environment. On-again, off-again operations of a factory or generating plant do not permit stabilization. If a particular segment of a body of water such as a cold-water stream were warmed 18°F for a long enough period, flora and fauna characteristic of the warmer water would develop, only to be wiped out at the first shutdown of operations, when the temperature would revert to the climatic norm.

Another problem is the effect of higher temperatures on oxygen absorption by water and its use by living organisms. The higher the temperature, the less oxygen the water will hold. Conversely, increased temperatures accelerate oxygen consumption in both respiration and oxidation on the order of about 200 percent (double) for every 18°F. At high temperatures accompanied by less oxygen in the water, organisms must be extremely well adapted (e.g., carp) or die from suffocation.

Some organisms cannot survive the shock of passing through the cooling systems of power-generating or industrial plants. How much their loss affects the ecosystem is somewhat proportional to the percentage of the total volume of water circulating through the system in a period of time. If the reservoir of unused water is large enough, the populations have a better chance of maintaining themselves at high levels.

Besides the thermal affects already mentioned, reproductive cycles, longevity, growth rates, and the relative proportion of primary food producers (primarily types of algae) may be altered. The primary producers of freshwater habitats are mostly diatoms, green algae, and blue–green algae. Of the three, consumers prefer diatoms most and blue–greens least. As the temperature increases, the number of species of diatoms decreases and the number of blue–greens increases. Between the two extremes are the greens. Since animals are selective in what they eat, any alteration of the food supply will change the relative size of their populations.

Some power companies term the heating of water **thermal enrichment** instead of thermal pollution. As has been indicated, heating does indeed favor some forms of life. A particularly good example of how heating affects populations is seen in the growth and reproduction of bass, turtles, and alligators in Parr Pond, which is heated by water discharged from a plant of the Atomic Energy Commission at the Savannah River Project near Aiken, South Carolina. These ectothermic (cold-blooded) animals grow considerably faster than would be normal because they can remain active throughout the year. Their foods seem to grow better also. If bass, turtles, and alligators were the only considerations, enrichment might be the proper term to describe the heating of water. But even if changes in the biotic composition caused by heating are somewhat undesirable, the loss must be weighed against human needs to decide what should be done.

Thermal pollution can be avoided or reduced by several methods. Among them are cooling towers (wet and dry), cooling ponds, and spray ponds. Where feasible, a more desirable practice would be to salvage the heat for some useful purpose like heating buildings, propagating fishes and shellfishes, irrigating crops in cool climates, or speeding bacterial action in sewerage systems.

REFERENCES AND FURTHER READINGS

Carson, Rachel. 1962. *Silent Spring*. Houghton Mifflin, Boston.

Gates, D. M. 1972. *Man and His Environment: Climate*. Harper & Row, New York.

Graham, F., Jr., 1966. *Disaster by Default: Politics and Water Pollution*. M. Evans, New York.

Hersh, S. M. 1969. *Chemical and Biological Warfare; America's Hidden Arsenal*. Doubleday, New York.

Kaill, W. M., and J. K. Frey. 1973. *Environments in Profile*. Canfield Press, San Francisco.

Leighton, P. A. 1961. *Photochemistry of Air Pollution*. Academic, New York.

Rudd, R. L. 1964. *Pesticides and the Living Landscape*. University of Wisconsin Press, Madison.

Stern, A. C., ed. 1968. *Air Pollution*. 2d ed., 3 vols. Academic, New York.

Wagner, R. H. 1971. *Environment and Man. Norton*, New York.

Wise, W. 1968. *Killer Smog: The World's Worst Air Pollution Disaster*. Rand McNally, Chicago.

Chapter 15 Solid Waste and Other Pollutants

PREVENTION AND PROTECTION

Pollution is largely a product of civilization, and unless controlled the problem grows along with the population. The more common pollutants are potentially more hazardous and difficult to manage now than they were in former times when they were less abundant. The old methods of disposal by diluting or burying that worked so well at one time are in most cases viewed as less desirable today. The problem also increases in complexity because mankind is constantly confronted with pollutants never before known. Many of these are either durable inorganics such as metals or synthetic organics like DDT that degrade slowly under natural conditions. To make matters worse, a whole new category of perceptual pollutants like noise and unsightliness offend the aesthetic sensitivity of man.

Solving the problems of pollution is a complex and expensive proposition. Sometimes the best solution is prevention. This is the thrust of some of the new laws that prohibit the release of pollutants into the environment. If pollution is unavoidable, then life must be protected from it. Usually protection is a matter of degree rather than one of elimination of the problem. This necessitates constant monitoring of the environment to guard against infractions that exceed acceptable standards set by some responsible authority.

SOLID WASTE

In the present state of technology, solid wastes are harder to dispose of than gases, solutes, and particles that can be swept away by air

FIGURE 15–1 *Air pollution.*

(Fig. 15–1) and water. Because they are bulky and often scattered, they are difficult to collect, transport, and dispose of, and this makes the entire operation expensive. A profitable salvage procedure is difficult because so much of the waste is a heterogeneous mixture that requires sorting. Incineration could be used for combustibles, but it has the undesirable side-effect of adding to air pollution, and dumping in landfills and elsewhere requires space and may pollute both air and water. Improved technology is sorely needed in solid-waste disposal.

In variety, solid wastes run the gamut from household refuse to industrial slag. Under present conditions some have value while others are worthless, and some are durable while others are easily destructible. Major categories of solid waste are:

Garbage
Paper products
Glass and ceramics
Metals
Wood
Plant residues (clippings, trimmings, straw,
stalks, sawdust, and shavings)
Textiles and leather
Plastics
Rubber products
Sewerage sludge and manures
Dead animals
Demolition debris
Cinders and ashes
Slag
Culm piles and mine tailings

Such an array of leftovers presents both problems of disposal and opportunities for secondary uses.

The cost of collecting and disposing of city waste varies according to disposal method, transportation cost, labor cost, and profits made from salvage or conversion products. At present the cost averages approximately $18 per ton. About 80 percent of the cost is for collecting. Litter collecting, by comparison, involves so much labor that the cost is about five times that amount.

The cost of disposing of solid wastes may be low, however, when one considers alternatives like grinding or pulverizing such wastes and attempting to dilute them in large volumes of water. The limits of sewage treatment facilities would be further strained, and cities downstream would have to install expensive facilities in order to have clean drinking water. Moreover, burying solids in landfills (Fig. 15–2) could be a way of accumulating valuable resources, awaiting technological advances whereby they can be recycled.

In landfill disposal a big cost factor is transportation, which makes it imperative that fills be located close to the cities. Cost can be further reduced by transporting larger loads in compactor trucks equipped with hydraulic presses that can reduce the volume of waste by about two-thirds.

The areas best suited for landfills have gullies, ravines, or slopes naturally contoured so that fill dirt is readily available for covering the waste. A bulldozer is used to shape the ground, move the waste

FIGURE 15–2 *A sanitary landfill. The bulldozer in the center of the deep trench is leveling and compacting the solid waste. When the trench is filled, it will be covered with soil.*

as necessary, compact it in the trenches, and cover it quickly. Quick burial is essential to keep trash from blowing away, to reduce the chance of fire, and to make the waste inaccessible to vermin. Fences should surround the fills to catch blow-away paper, and fires should be avoided so as to prevent air contamination.

When a landfill is full, its best potential is for recreational or park uses. It is not recommended for construction sites because the earth settles and sometimes tilts buildings or cracks their walls. It has also been reported that there is danger from toxic and explosive gases generated by anaerobic bacterial activity in the organic materials beneath the surface. These materials can also be a serious source of water contamination.

In addition to innumerable trash burners operated by individual homeowners and small businesses, some cities dispose of solid waste by incineration. Many of the facilities are substandard and allow too much residual matter to escape into the atmosphere. Incinerators can be built close to the source of the waste, substantially reducing the cost of transportation. About 75 percent of the collected material can be burned and used to generate electricity. By burning under pressure, the combustible solids produce hot gases capable of powering turbines. Although they are only about one-third as good a fuel as high-grade coal, these unlikely combustibles can operate a generating unit capable of supplying about 10 percent of the electricity needed by the town that discards them. The remaining noncombustible glass, metal, ashes, and plastic must be disposed of in some other manner.

A few cities have built **composting** facilities. After the removal of inorganic substances such as glass and metals, the remaining materials are degraded by bacteria into a peatlike organic compost suitable for fertilizing and conditioning soils. The waste is first pulverized to provide maximum surface area for bacterial action, and then the proper amount of aeration and humidity is provided to ensure rapid bacterial growth.

Another possibility for disposing of solid waste is converting it into useful products. The Bureau of Mines is converting waste by a thermal process known as destructive distillation. In some places the garbage portion of solid waste is fed to hogs and thus transformed into meat. To avoid trichinosis and other diseases, law requires that the garbage be pasteurized.

Some coastal cities have taken the easy way and dumped waste into the ocean. The consequences of that kind of disposal are under study, and the practice is likely to encounter more and more opposition as time passes. Problems already encountered include contamination of shellfish waters and swimming areas.

A particularly thorny problem is the disposal of worn-out cars, of which an estimated 11–15 million are retired or abandoned annually. Not only are they unsightly (Fig. 15–3), but they are expensive to handle and have little salvage value. When salvaged, the cars are stripped of usable parts and the remaining metals are reused. The ferrous scrap can be mixed in varying amounts with ore, depending on the product to be manufactured.

Several techniques for salvaging cars have been tried. Often they are burned to remove all but the metal carcasses, which are then compacted in large presses to facilitate handling. The most recent and promising technique is shredding and electromagnetic separation. More sophisticated methods of separation are under study.

FOOD CONTAMINATION

Food products contain many things in addition to energy-producing proteins, lipids, and carbohydrates. Among these are natural products that may or may not be harmful and other substances added deliberately or accidentally. Some of the ingredients may be poisonous to anybody who consumes them; others may be harmless in

FIGURE 15–3 *Junked cars dumped on the banks of a waterway.*

small quantities but deadly in large quantities. Some are known to be useful in trace amounts and lethal in large amounts. The effects, if any, of many are still unknown. Human responses are so different that some of these substances may harm one person but not another. All of these variables must be taken into account before a food can be declared polluted.

Spoilage may make foods unsightly, unpalatable, or dangerous. It is caused by internal chemical changes, as when a fatty substance becomes rancid, or by the metabolic action of bacteria. Some bacteria excrete toxins. A particularly deadly bacterium is *Clostridium botulinum*, which causes the poisoning known as botulism. This organism produces toxins under anaerobic conditions before the food is consumed. Other types of dangerous bacteria are ingested with food and multiply rapidly in the digestive tract, where they produce toxins. Examples of this type are in the genus *Salmonella*.

Some plants and animals are naturally poisonous but seldom cause trouble because they are avoided by man. When they are consumed, it is usually by accident. They may be mistaken for a suitable food, as sometimes occurs with poisonous mushrooms. Children are particularly prone to eat unsuitable foods. Of considerable historical interest is the parasite ergot (*Claviceps purpurea*), a poisonous fungus that grows on rye and other grains. When infected grains are used for food and consumed in sufficient quantities, severe symptoms occur, occasionally leading to death. Some of the symptoms are similar to those of lysergic acid diethylamide (LSD) poisoning. If consumed repeatedly, ergot interferes with circulation by causing blood vessels to contract. This often results in a gangrenous condition followed by the loss of digits, hands, feet, or limbs. During the Middle Ages, before the cause of ergotism was known, it was a scourge among peasants who depended on rye bread as the main staple of their diet. It also was a bane to cattle. In this century cases of ergot poisoning have occurred only sporadically.

From time to time poisons accidentally get into foods. With the use of so many insecticides, herbicides, and fungicides, the chance of accidental poisoning is much greater than formerly. In some cases, as with DDT or mercury, the poisons are incorporated in the food product itself. Like the two just mentioned, radioactive fallout is absorbed and incorporated in human foods. This is one reason for concern about atomic explosions.

Not all of the food contaminants are necessarily poisonous, although they may be undesirable. Cow's milk may have a bitter taste if the cows eat certain plants or an onion taste if they eat wild onions. Even meats pick up undesirable flavors such as the fishy taste of pork caused by feeding fish scraps to hogs. A dangerous practice that has been discontinued is stimulation of growth in cattle and chickens by feeding them a female sex hormone. Entirely too much of the hormone was passed on to humans in milk and meat.

Over the years a number of **additives** have been used in foods to serve specific purposes. Some of these are salt, sweeteners, dyes, flavors, thickeners, emulsifiers, buffers, vitamins, minerals, bleaches, and preservatives (antioxidants and biological inhibitors). Such substances could cause trouble to individuals in cases where they are allergic or have unusual physiological conditions. The practice is to regulate the amount used and to keep the quantity at "safe" levels.

Sodium nitrate is often used to fix color in meats so they will appear fresh and to inhibit organisms that might cause spoilage. Farmers have used the same chemical (nitrate of soda) for fertilizer, and it has killed farm animals that accidentally consumed it. Under anaerobic conditions in the digestive tracts of animals, bacteria reduce nitrate to nitrite, a form that readily combines with hemoglobin and thereby reduces the capability of the blood to transport oxygen. The strategy is to use just enough in foods to reduce the activity of the decay organisms without damaging human cells. Whether this can actually be done is questionable. There may be no visible or measurable damage, but this does not mean there is no harm at all. The same danger exists in many other food preservatives. People should be aware of the possibility that potential poisons to humans may be masquerading under the name of preservative.

The Food and Drug Administration is the governmental agency responsible for approving food additives. The number of approved additives in use is well over 2000. Safety tests are run on laboratory animals, but not everyone is convinced that such tests are entirely dependable. For one thing, humans are not exactly like the test animals. Then there is the possibility that an additive may be a slow poison that requires time to express itself or may have a latent effect not obviously associated with the cause. Many additives, like cyclamates, were once accepted as safe but were taken off the market when problems arose.

Adulterated foods are particularly offensive to the consumer because of the peril to him and the deceitful marketing practices. Concealing the fact that a meat is spoiled by masking odors or color with chemicals is a dangerous practice, much more serious than putting bad fruit in the bottom of the basket. Among many such deceitful and dangerous practices is that of substituting mineral oil for cooking oil. In some cases, however, adulterants may win a measure of public acceptance, as in the case of chicory in coffee.

There is increasing concern that contaminants in food and water may be worse than supposed. There is some evidence that they may cause genetic changes, birth defects, cancer, and circulatory diseases. It is sufficiently adverse to justify greater caution in the testing and approving of chemicals that may be consumed by humans. Moreover, much more information must be obtained on what happens to various substances within the body, because some seemingly harmless ones could be changed into deadly products as they are metabolized.

RADIOACTIVE POLLUTION

Among the elements of the universe there are some unstable ones that emit harmful radiations. Life is constantly exposed to such radioactive emissions but is located at the surface of the earth where exposure is minimal. Radiation that constantly comes from outer space (cosmic radiation) and from the earth itself is known as **background radiation**, as distinct from radiations caused by man. The latter come from bombs, nuclear reactors, cathode-ray tubes (e.g., X-ray and television picture tubes), electronic devices, and radioactive isotopes used in medicine and science.

Radiations may be classified as **ionizing** and **nonionizing** according to whether or not their energy disrupts the energy balance in otherwise stable compounds. If they are ionizing radiations, they knock electrons from atoms or compounds, leaving them with a positive electrical charge. Chemical bonds may be broken, causing the molecules to separate into ions (electrically charged particles) or free radicals, which can then react with other molecules within the cells. The altered molecules (e.g., DNA), no longer able to function as before, cannot serve the cells as they should, and the cells die if the damage is great enough.

The three major types of radiation, **alpha**, **beta**, and **gamma**, are of the ionizing type, as is X radiation, which is similar to gamma radiation. The nonionizing types that come from electronic devices like lasers, microwave ovens, ultrasonic vibrators, and switching and communications equipment are as a whole not as hazardous as the ionizing types.

All of the ionizing radiations damage cells, but the amount of damage is related to their ability to penetrate the body. Alpha particles, the positively charged particles having two protons and two neutrons, can do little harm because they can hardly get beyond the epidermis. Beta particles, the negatively charged electrons, can penetrate a thick sheet of cardboard or as much as a half inch of some other substances, depending on their density. They can "burn" the skin, although clothing is a good shield against both beta and alpha particles. Greater damage can be done internally if alpha- and beta-emitting matter is accidentally inhaled or ingested. Gamma and X rays are more energetic and consequently are deeper penetrators. While they easily pass through both clothing and the body, they can be reduced to relatively safe levels by several inches of lead or several feet of concrete.

Radiations are measured in terms of **curies** (c) or **roentgens** (r). The curie is a measure of the amount of decay of a radioactive substance in a specified period of time, and the roentgen is a measure of dosage, the amount of radiation required to cause a certain number of ionizations in a specific volume of air. Other units of measurement have been devised to express specific aspects of radiation effects. One is the **rem** (roentgen equivalent man), or the amount of any type of radiation that produces a biological effect equal to the absorption of one roentgen of X radiation.

The amount of radiation a human can stand depends on a number of factors such as type of tissue, age, condition of health, and size of area exposed. A dose (single exposure) of 1000r will definitely kill. A dose of 500 r, half that amount, will kill 50 percent of the people and is usually expressed as LD-50 (lethal dose for 50 percent of the population). A dose of 100–300 r can usually be overcome but often leaves a person permanently damaged. The tissues most sensitive to radiation are those that actively divide, such as blood-forming tissues (marrow, lymph nodes, and young spleens), some epithelium, and gonads. Embryos, fetuses, and children suffer more than adults because a larger portion of their cells are actively dividing.

Survivors of heavy radiation are faced with some threatening prospects. They may quickly lose their hair or develop cataracts, and they may have insidious internal changes not unlike a time bomb waiting to explode. Longevity is reduced in mice, and this may also be the case in humans. It is a well-known fact that cancer and genetic changes can occur. The latter are of the greatest consequence to future generations, to which the defects are passed.

Radiation pollution has the special problems of often persisting for long periods and of not being detectable by the natural senses. The longevity of radioactive substances is measured in terms of **half-life**, or the time required for half of it to decay. In the same additional length of time half of the remaining portion decays, and so on ad infinitum. The half-lives of some of the more familiar isotopes are:

Bromine-85—3 minutes
Calcium-45—164 days
Carbon-14—5600 years
Cesium-137—30 years
Cobalt-60—5.27 years
Gold-198—64.8 hours
Hydrogen-3 (tritium)—12.26 years
Iodine-131—8 days
Iron-59—45.1 days
Krypton-85—10.76 years
Nitrogen-16—7.35 seconds
Phosphorus-32—14.3 days
Potassium-40—1.3 billion years
Radium-226—1620 years
Sodium-24—14.8 hours
Strontium-90—28 years
Sulfur-35—87.1 days
Thorium-232—10 billion years

Some radioactive materials have an affinity for substances in living organisms and become part of the food chain. Cesium-137 behaves physiologically like potassium, which is concentrated in muscles, and is therefore found in meats.

Strontium-90 is accumulated by plants and passed on to the animals that eat them; humans are likely to get it from cereals and milk.

Like calcium it becomes concentrated in bone tissue. Cobalt-60 and gold-198 tend to accumulate in cancerous cells, and iodine-131 concentrates in thyroid tissue. Zinc-65 is readily absorbed by some diatoms, oysters, clams, and scallops. In a pond at the Savannah River plant near Aiken, South Carolina, algae were found to contain zinc-65 in concentrations up to 6000 times its quantity in the environment, and bluegills concentrated it in their bones by a factor of 8200. Other radioactive materials, like krypton-85, do not enter metabolic pathways and do not form part of the food chain.

Radiations from isotopes have some important uses. Biologists use them as tracers to study transport and metabolic pathways. Medicine uses them for the same purposes, to photograph internal structures, and in cancer therapy. Industry uses radiations to detect leaks in pipes, to monitor the flow of liquids, to measure thickness, to measure wear on moving parts, and to sterilize foods and pharmaceutical products. They may also be used to trace the patterns of air and water currents. Geneticists use radiation to cause mutations, some of which improve plants and animals. Archeologists and earth scientists use them to determine the age of artifacts and formations. At present one of their fastest-growing uses is as a fuel to generate power.

To meet the energy requirement today and in the future, utilities are switching to the use of electric generators powered by radioactive fuels. The Atomic Energy Commission predicts that 600 breeder reactors will be in operation by the year 2000. However, the problem of disposing of plutonium and other solid waste and the release of tritium into the atmosphere from these reactors are causing some anxiety.

How can the operators of nuclear power plants dispose of so much poisonous waste? Shall it be buried on earth or shot by rocket into the sun, which is already seething with radioactivity? The latter method is currently uneconomical. At present companies are allowed to operate processes that leak small amounts of radioactive substances into the ecosystem. Liquid and solid radioactive wastes must be packaged and buried, and where feasible, wastes with highest activity are processed for reuse.

Underground disposal of radioactive chemicals is becoming increasingly difficult as local citizens resist the use of their neighborhoods for such purposes. Recently a scheme to discard such chemicals in an old salt mine in the Middle West was fought because of the fear of possible leaks. A similar plan to bury radioactive waste on the grounds of the Savannah River Project is meeting with increasing opposition because the site is located in a high-risk earthquake zone.

In 1969 the total amount of radioactive solid waste was 1,800,000 cubic feet. Unless there is an unexpected reversal of trends, there will be a phenomenal growth in the use of reactor-driven generators accompanied by a corresponding mountain of waste. Large quantities of liquid wastes now stored in tanks are being converted into more compact solid form for burial. Unfortunately, the wastes will not go away but will remain buried and out of sight. No matter how well the containers are constructed, they are not permanent. They only delay the release of radioactivity into the ecosystem. Catastrophes and inevitable deterioration of the containers will eventually set the dangerous chemicals free. The buried waste is like a booby trap waiting to be sprung. The only offsetting factor that slightly mitigates the bleak outlook caused by having so much waste in storage is the gradual decline in radioactivity due to the natural decay of radioactive substances.

Tailings from the mining of radioactive minerals are another hazardous form of waste. Erosion disperses some of the radioactivity so that the water of some streams is definitely dangerous. Such tailings were recently used by housing contractors as a base for slab floors and backfilling for basements. When the mistake was discovered, some houses were abandoned because of the danger.

VISUAL POLLUTION

Visual pollution consists of trash scattered about (Fig. 15–4), inscriptions on rocks or water tanks, names carved on trees, paths across lawns, dilapidated buildings, cluttered billboards, gaudy neon signs, destroyed forests, eroded ground, and the like. It does not speak well for the aesthetic sensitivity of man and is indicative of a quality of life far beneath the aspirations of the best of mankind.

To know what visual pollution is, one must first know what is beautiful. Probably most people agree that order looks better than disorder, that cleanliness looks better than filth, and that a natural landscape looks better than buildings and pavement. Unfortunately, some people are conditioned to the worst and know no standard with which they can compare their ugly surroundings. Others know better but lack personal or public pride.

So many people are offended by the sheer ugliness of their surroundings that they are beginning to react against those who would despoil the environment. As a consequence of this concern, redwoods are being saved and wilderness areas set aside for the protection of wildlife. In the cities slum clearance is progressing, and planning departments are suggesting revitalization projects. Real-estate developers are beginning to save trees and blend houses into natural settings. Electric and telephone companies are cooperating by burying unsightly transmission lines.

FIGURE 15–4 *Visual pollution.*

Americans would do well to follow the practice of the Swiss, who before building a house are required by law to erect a pole as tall as the proposed building on the site where it will be built. This practice gives the public a chance to see if the building will obstruct a view it wants to preserve and allows time for any interested person to block the construction by legal means if he so desires.

Where voluntary concern for the public interest is ignored, legal action is taken more frequently. For instance, operators of strip mines are being forced to cover the scarred surface of the earth with plantings. Although much still needs to be done about junkyards, some of the owners have been required to shield them from view by planting hedges or by erecting fences. Along some highways advertisers have been required to eliminate roadside signs, and further regulations are proposed.

NOISE POLLUTION

Sound is an important sensory contact with the environment and is obviously needed and enjoyed. But some sound is nothing more than noise and as such can be designated as pollution. By this definition, noise is the portion of sound that is unwanted because it is either distracting, irritating, or harmful. Here the human variable is involved, for some might classify the deafening "hard rock" sound as music while others might classify it as noise. Since amplified "hard rock" is known to damage the body, it seems appropriate to call it noise in cases in which damage is done.

Urban dwellers and certain occupational employees are most concerned about noise simply because they are subject to more of it (Table 15–1) than is normal. Industrial operations, construction, and transportation usually account for most of it. Some of the most violent noises are made by jet planes. The sonic boom caused by a jet travelling faster than sound is enough to rattle houses and break windows. In some places high noise levels are incessant, and relief of a sort comes only during sleep when the brain fails to interpret it. Even then the subconscious may not turn it off: It is known that exposure to noise during sleep can cause measurable physiological changes in the body.

The dissimilar ways in which humans react to sound make it difficult to appraise all of the possible damages to mind and body. Clear-cut cases of hearing impairment resulting from persistent loud noise are the best documented. Long before this generation, hearing losses were common among coppersmiths, blacksmiths, and boilermakers who were exposed to hammering noises. Noise above the ninety-decibel level for prolonged periods is known to cause permanent damage to the ears. Of course the level of a noise is related to the distance of the listener from its source. Noises that may register above the danger level are made by heavy trucks, jet planes, riveting machines, automobile horns, shouting humans, bands, sirens, subway trains, and hammering on metal. Besides impairing hearing, noise may also cause vasoconstriction accompanied by a variety of related effects. Long exposure to noise may lead to hypertension or ulcers, especially if the noise is irritating.

Some effects of sound on other animals have been observed from time to time. Soothing music apparently evokes favorable reactions in some domesticated animals, whereas noise obviously startles them. In one case poultry producers in an area where army maneuvers were in progress noticed a drop in egg production. They attributed the decline to the noisy helicopters, jets, and trucks that interrupted the peace to which the birds were accustomed.

What can be done about noise? Technology can no doubt produce quieter machines and processes. Better building materials and wall construction to insulate against sound could confine or exclude most of it. If noise goes uncontrolled, the individual can resort to ear protectors to muffle the sounds.

TABLE 15–1 *Sound pressure levels of common sounds and noises*[a]

Sound quality	Decibels	Sound source
		rocket engine
threshold of feeling/pain		ram jet
	120	turbojet: 7000 pounds thrust
		propeller aircraft
		boiler factory
deafening	110	nearby riveter, drop hammer
		thunder
	100	subway and elevated trains
		woodsaw, punch press
	90	loud street noises
very loud		noisy factory, screw machine
		pneumatic drill
	80	police whistle, portable sander
		noisy office
loud		average traffic
	70	normal radio
		average factory
	60	
		noisy home
moderate	50	average office
		ordinary conversation
		quiet radio
	40	
		quiet home
faint	30	private office
		average auditorium
	20	quiet conversation
very faint—threshold of audibility	10	rustle of leaves
		whisper
	0	soundproof room

[a]Source: Theodore Berland, 1970, *Medical and Legal Consequences of Noise Pollution,* AMF Beaird, Inc., Shreveport, La.

THE POLLUTION OF WARFARE

Individual wars come and go, but strife seems to be as permanent a part of the environment as anything else. The polluting effects of war are obvious. Explosives, fires, herbicides, and machines destroy the stability of natural communities by killing outright, contaminating the air and water, and accelerating erosion. The more appallingly poisonous chemicals as well as biological and nuclear weapons have been more or less kept under wraps. What is difficult to evaluate and seldom mentioned is the emotional strain war inflicts on humans through fear, hatred, grief, and despair.

Even nations far from the battlefield feel compelled to test weapons and go through mock maneuvers (war games) that damage the ecosystem extensively. The testing of atomic explosives in remote areas is an example. If the great powers cannot agree to stop the testing at a time when only a few nations have the technology to develop atomic devices, how can one hope for easier times in the future when other nations achieve the know-how and potential? With the proliferation of nuclear power stations of the breeder type, the possibility exists for a small nation to construct nuclear weapons from the plutonium it generates and to use them to overpower or blackmail other countries. As for the war games, the waste of resources such as petroleum is matched only by the devastation of the countryside. No wonder the local citizenry often objects to the maneuvers, even when performed on public lands.

The problem of disposing of unwanted or obsolete but dangerous materials in the national arsenal is a by-product of the preparations for war. The stockpile includes among other things germs and poisonous gases. The Army's planned and eventual disposal of nerve gas in the Atlantic hit the headlines in 1971. Because the containers were old, people feared that they could not be transported safely or that once dumped they might leak rapidly and cause local damage. The presumption that even very high dilutions of these substances in the ocean might be harmless is open to question. The possible consequences of the plan were enough to draw the verbal fire of other nations.

REFERENCES AND FURTHER READINGS

Burns, W. 1969. *Noise and Man.* Lippincott, Philadelphia.

Commoner, B. 1967. *Science and Survival.* Viking, New York.

Johnson, H. D. 1970. *No Deposit—No Return; Man and His Environment: A View Toward Survival.* Addison-Wesley, Reading, Mass.

Novick, S. 1969. *The Careless Atom.* Houghton Mifflin, Boston.

Schultz, V., and A. W. Klement, Jr., eds. 1963. *Radioecology.* Reinhold, New York.

The First Annual Report of the Council on Environmental Quality. 1970. *Environmental Quality.* GPO, Washington, D.C.

Chapter 16
The Future: A Perspective

CRISIS OR OPPORTUNITY?

There are optimists and pessimists on every question. All acknowledge that the quality of the environment is a problem, but opinions about its seriousness vary considerably. The more emotional critics are talking crisis and catastrophe, while the more temperate ones, who may be no less concerned, do not believe a calamity is imminent. Rather, they think there is plenty of time to look at all aspects of the problem in a deliberate manner, taking into consideration the disruptive and expensive consequences of acting precipitously. The danger is that the former may lose their ardor while the latter drag their feet.

Man's greatest mistake has been and continues to be his insensitivity to his environment. By his actions he seems to be saying that he has the world in his hands instead of recognizing that he is a part of it. He selfishly takes what he wants and discards what he does not want, disregarding almost entirely his contemporaries and descendants. For the most part he makes little effort to preserve the natural environment or to harmonize with it.

Hope for the future lies in nature's own restorative processes and to a lesser extent in mankind's accumulated wisdom. Scars in the landscape caused by man or by natural catastrophes are quickly repaired as organisms and environment work together to bind and cover them. Fortunately, some depleted resources are renewable because the primary energy input is still coming from the sun. With such dependable natural allies, man should be able to repair much of the damage he has done. Surely he can find substitutes for exhausted and irreplaceable resources, conserve what is left, reclaim damaged habitats, and even improve his surroundings according to his desires and needs. To expect less of him is to underrate him.

Man has learned how to extract all that he needs from the natural environment. As his needs or requirements have continued to rise at ever-increasing rates, he has had to further short-circuit or disrupt the system. Although he has been able to "buy time," the magnitude of these disruptions is such that his own survival is now threatened. Survivorship is the only criterion of success for a species in the biological world—one mistake and the species could become a fossil. Now is the time for the human species to develop an ecological strategy that will serve not only the present generation but also those of the future.

It is necessary to develop a program of **ecological maintenance**, which is to quality living what preventive medicine is to health. There is too much risk involved to postpone giving attention to it or to leave it to chance. The time to begin is now, because developing proper policy will require a great deal of time and patience. For many of the problems there will be honest but conflicting ideas about solutions. Not until a substantial amount of research is completed can such differences be settled. The greatest mistake would be to expect instant answers and total success; the greatest success would be to view the problem as a continuous one and develop long-range policies.

FINDING OUT WHAT IS NATURAL

Undisturbed communities and environments are already difficult to find in many parts of the world. As a result of human activity, only fragments of climax communities remain. A virgin stand of timber, a patch of prairie, or a wilderness attracts environmental scientists the way nectar attracts bees. Environmentalists generally try to get these areas set aside and often resist opening them to the public. The motive to keep such places natural is to maintain a **standard of reference**.

The only way to appraise the damage suffered by natural areas is to contrast their present condition with their natural condition in the past. Since no one now alive saw much of the United States in a natural condition and few scientific studies were made of the environment in those earlier days, there is no way of knowing how much harm has been done. Although many people have seen the destruction of vegetation, the spoiling of streams, and the fouling of the air, specialists need to measure the amount of change and to set ranges of tolerance for the survivors.

It is important that some of the preserved areas be large. They may never again be large enough to influence climate, but they should be large enough to safeguard animals that require extensive home ranges. Some species are known to require several square miles per individual. The matter is further complicated by animals' disregard for boundaries. They are often well protected when they remain on a game perserve but slaughtered as soon as they cross onto private land.

Natural areas can and should be established to serve the scientific community for study and the public for recreation. These needs may conflict, but the compromise may be worthwhile if the public can be better educated to the realities of nature by having a place to see them relatively undisturbed. With heavy human traffic, natural areas will not be completely undisturbed, yet they can be maintained in nearly natural condition if they are properly managed.

CULTURAL CHANGES

The paths, trails, burrows, wallows, and other imprints made by wild animals are insignificant compared to the shambles left by the activities of man. The changes he has wrought are called cultural features. No one would question the need to alter the ecosystem to serve human needs, but one might seriously object to the methods used. Cultural changes should not exceed what is needed nor be made before studies virtually ensure that there will be no adverse ecological consequences.

An indiscriminate policy of altering natural watercourses can be costly and, worse still, can have drastic effects. To suppose that the draining of a marsh to produce dry land or the dredging of a channel to enhance navigation always improves the environment is a simplistic conclusion. The zeal of local developers usually blinds them to every consequence except the economic one from which they hope to profit. Thus, to the consternation of its backers, construction of the Cross-Florida Barge Canal was halted when a few self-appointed protectors of the environment fought the project in court.

More and more, government agencies have become protectionist advocates where large changes are proposed. Such was the case when the South Carolina Water Resource Commission (1971) found evidence that further dredging of Port Royal Sound near Beaufort should be avoided or there would be serious adverse effects on underlying freshwater strata, which in turn would damage adjacent land areas. A suitable alternate channel may eventually be found, although some possibilities have already been rejected as unsatisfactory. In many coastal areas care must be taken not to charge underground aquifers with salt water.

A number of possible disadvantages, some serious, must be anticipated before changing natural drainage patterns. No longer is the Army Corps of Engineers' practice of channelizing streams accepted without question. In fact, some local groups have flatly rejected proposals to channelize their streams. Probably it is inadvisable to make even such minor changes as clearing stream channels of trees that fall into them for the simple reason that the small amount of clogging retards the flow, favors the retention of groundwater, and provides backwater habitats. Drainage of standing water lowers the water table and destroys the aquatic habitat, which among other things might be used by migratory waterfowl or for recreational fishing. Tampering with water levels seems to pose the greatest threat to underground water supplies. Of course, the impounding of reservoirs has the opposite effect of increasing underground water reserves.

Pavement and buildings cover land that might otherwise be occupied by vegetation and the accompanying animal life. They reduce the ground surface into which rainfall seeps and hasten runoff. In the country, where most of the surface is vegetated, the plants exert a moderating influence on the temperature of both the soil and the lower atmosphere.

Other problems, as yet unassessed, are the dangers to wildlife caused by miles of highways. The number of small animals killed by automobiles is considerable, and fenced highways interfere with the free movement of animals. Flying animals and some small ground animals are unimpeded, but large ones are isolated and cannot freely interbreed or search for food. Many rights-of-way are sprayed with herbicides (Fig. 16–1) to control the growth of weeds and trees. Alteration of the floral composition, as well as the presence of unnatural chemicals, could be detrimental to some animals.

MAJOR PROBLEMS

Perhaps the root of most environmental problems is population pressure. The term **carrying capacity** expresses the fundamental concept that a particular area is capable of supporting a specific population size. Although the term is used primarily in connection with wildlife, the same principle applies to the earth and its capacity to support a human population. Some parts of the earth have already exceeded their optimum capacity, and the result has been starvation and misery. Although no one can give an exact figure on how many people the ecosystem can support, the prospect of a doubled population at some future date is frightening. It is difficult to envision how a proper proportion of natural environments can be maintained while simultaneously supporting such masses of humans.

FIGURE 16–1 *Spraying plants with a weed killer along a roadside in the upper Michigan Peninsula.*

Large populations are heavy consumers not only of foods but also of many other resources. Some resources are already scarce, and others cannot last long at the present rate of consumption. The alternatives are clear—free or restricted use of resources. Those who argue for free use claim that substitutes can be found when the need arises. Such an attitude is like that of the man stranded on a desert with a canteen of water who decides to drink it all at once, leaving his future to fate. The second alternative is to make resources last as long as possible on the chance that substitutes might not be found quickly.

Conservation of resources is not simply a matter of rationing, although that may be a worthwhile objective. Resources must be put to repeated use. Water, for example, can be used many times—for municipal water systems, generating electricity, irrigation, and sports or navigation—as long as it is properly managed.

One resource in particular needs to be conserved: phosphorus, an element essential to life. It is an important constituent of nucleic acids and the energy-rich compound adenosine triphosphate (ATP). Phosphorus, nitrogen, and potassium are the maor ingredients of fertilizers because they are frequently limiting factors in agricultural soils. Of the three, phosphorus is the least abundant in natural reserves available for use as fertilizer. It is unfortunate that existing deposits of phosphorus are being rapidly depleted and ultimately dispersed as waste in the depths of the ocean, where dilution and sedimentation make reclamation all but impossible.

Phosphorous lost from fields and sewage systems contributes to the problem of eutrophication. The removal of phosphorus before it enriches aquatic environments is not widely practiced, although it should be. Phosphorus is removable by chemical precipitation techniques that are already economically feasible. Coagulants such as ferrous chloride, ferrous sulfate, ferric chloride, ferric sulfate, aluminum sulfate, sodium aluminate, and lime (Ca) can be added as a step in the treatment process. The free ions from the additives combine with the phosphates to form a precipitate that is easily separated from the liquids of the system. About 80 percent of the phosphates can be removed by the usual settling process and an additional 10 percent by a postsedimentation filter. Some industrial chemists see pollution control of this sort as more than an environmental necessity; an economic profit could be derived from the recovered products.

It should be noted that phosphorus and nitrogen, although maligned for causing eutrophication, may sometimes have to be added to disposal systems where these substances are not present in the waste. The reason is that most disposal systems depend heavily on biological organisms like algae and bacteria to reduce organic waste directly or indirectly. The organisms must have both elements in order to flourish.

Besides consuming vast quantities of natural resources, man is a heavy consumer of energy, which he uses for heating, cooling, and running machinery. The differences in consumption of energy by different peoples of the world are enormous. While millions of Asiatics save animal dung for fuel to cook their foods, Americans use atomic energy to air condition homes. Like primitive peoples of earlier generations, many people still rely largely on combustible materials such as wood, while technological societies have turned to the fossil fuels—coal, oil, and gas. Along with the use of these fuels has come an increase in pollution. Now the need has arisen for new energy sources, partly because the fossil fuels will not last indefinitely and partly because of the problems of pollution.

Possible major sources of energy for the future are sunlight, wind, ocean waves, and atoms. The impending shift is clearly from fossil fuels to atomic fuels. According to the United States Department of the Interior figures (Table 16–1), present and predicted changes will increase the use of atomic energy from about 0.3 percent now to 22.7 percent by the end of the century. The fast breeder reactors will probably be in use within the next ten years and are expected

TABLE 16–1 *Anticipated changes in use of energy sources in the United States*[a]

Energy Source	Year 1971	Year 2000
oil	43.0%	34.6%
natural gas	32.8	26.4
coal	20.1	13.7
waterpower	3.8	2.6
atom	0.3	22.7

[a]Source: U.S. Department of the Interior.

to increase the efficiency of uranium fuel by at least thirty times. In the future, moreover, harnessed atomic fusion may become a reality.

The transition to a cleaner source of fuel will not be sudden. For one thing, demand for energy is expected to increase by 50 percent or more during the next decade, and the technology of cleaner sources is not fully developed. Meanwhile, the search for oil continues. Ultimately oil may be extracted in commercial quantities from large oil shale deposits like the Green River shale formation of Colorado, Wyoming, and Utah, where the reserve is estimated in trillions of gallons. Further antipollution measures may be taken to reduce noxious emissions such as stack gases from the combustion of coal.

A large problem in developing and achieving cooperation for an environmental strategy at the political level arises from the magnitude of many problems that cannot be handled by individuals. Politicians are often unjustly criticized by their constituents, who are much less informed and often highly emotional, for inaction or for making unpopular decisions. There is absolutely no reason to think they are indifferent to the problems that confront the nation. They must rely on "expert" opinions that are often contradictory, but as answers become clear, correct actions are likely to be made public policy.

Not the least of environmental problems is lack of personal discipline in the utilization of energy and other resources. Complaints about pollution caused by the automobile engine have a ring of insincerity when they come from families with a car for every member, some of whom spend their idle hours cruising the streets in cars and tearing up the countryside with dune buggies.

MAINTAINING A QUALITY ENVIRONMENT

A great danger is that the degradation of the environment will be so gradual that bad conditions come to be accepted as normal. Consequently, each succeeding generation is likely to accept a lower standard than the previous one.

What is a quality environment? Merely what man is accustomed to? Certainly the desired environment should at least be capable of supplying basic human needs in abundance as well as safe, uncongested, and possessing a quality of naturalness.

One of the greatest blights, of course, is the urban decay due to overpopulation. High population density with all of its problems, plus brick and concrete, have all but destroyed the balance and orderliness of nature. Even city parks are so abused that much of the cost of maintenance must go for cleanup and repair.

The greatest obstacle to urban progress is lack of individual responsibility, cooperativeness, and community pride. The first concrete step is to refrain from littering and to instigate periodic cleanup campaigns. If the aimless and destructive energy of many urban dwellers were redirected to constructive ends, the transformation would be startling.

Urban living could also be improved by such things as efficient, fast public conveyances. Staggered work shifts that bring employees to work and let them out at different times could be scheduled, and new industry could locate close to its labor source.

Whether the environment is changed for better or worse will depend on the size of the population and what is needed to support it. The need for greater productivity to support overpopulation will result in a larger portion of the earth's surface being put to monocultural crops, which are distinctly more productive in terms of human needs than natural communities are. Just because natural communities are less productive in this sense does not mean, however, that they have less value in the ecosystem. They play fundamental roles in the balance of nature in addition to their more visible value as landscape and their acknowledged usefulness as refuges for animal life.

INDIVIDUAL FREEDOM, OWNERSHIP, AND GROUP RIGHTS

All organisms require space to live, some more than others. Some animals intermingle with a minimum of friction simply because they occupy different niches and do not compete with one another. Others claim territories, which they defend vigorously. Some humans seem to need isolation and acquire space very much the way animals claim territories. Through the rights of ownership they can hold property for themselves and exclude others. There is some question, however, about where an individual's rights begin and end. It is in essence a problem of individuality versus the society of which the individual is a part.

Although a plot of land may be privately owned, misuse of that plot may concern other people. The air that traverses it, the water that drains from it, and the noises, odors, and sights that emanate from it affect others. The owner cannot claim what he cannot control, and what he cannot control he has no right to originate or discard. What the individual cannot claim is public property and must be respected as such. Not only should the owner protect the public from anything offensive that might come from his property, but he should not be granted unlimited rights to obstruct views.

To many people water is almost as free as the air, yet water is generally allocated by law in arid lands. A person who owns land is not automatically entitled to dig a well, dam a stream, or divert water for irrigation. Water rights are more clearly defined in some states than in others.

Wildlife that knows no bounds is clearly public property and must be protected by law from greedy or malicious people. Ordinarily, the average citizen welcomes such regulations protecting his own share of this natural resource.

In no society is the individual given total freedom. But in compensation for this loss he receives benefits that only society can provide out of the pooled knowledge and resources at its disposal.

NATIONAL AND INTERNATIONAL ACTION

As a macroecosystem, the world is a cohesive unit with interlocking influences. Superimposing political boundaries upon it does not alter those influences and does little more than restrict the free movement of humans. Other organisms or their disseminules (e.g., seeds and spores) cannot be similarly restricted; neither can flowing air or water be substantially altered from their courses. As they sweep from one political unit to another, they pick up, transport, and deposit their loads of impurities. If these deposits are undesirable, concern about their origins is a natural response.

A river can be a source of interaction between several places, one after another, from its headwaters to its mouth. Rivers like the Ohio and Mississippi furnish water for a city only to have much of it returned through its sewerage system. Downstream, another city uses the same water and returns it to the stream again, and other cities follow suit. But the stream is common property and must be protected from misuse. Just as one city should not be allowed to pollute a stream serving others downstream from it, one state or country should not be allowed to pollute water that passes beyond its border. Nor should upstream users divert too much water or interrupt its constant, even flow. If a river forms a boundary between two countries, neither should foul the water or divert a disproportionate share for its own use.

By virtue of their locations, interstate and international rivers require cooperative management to avoid ill will between their users. Aside from the almost universal problem of pollution, a number of other difficulties arise from time to time. Both the St. Lawrence and the Rio Grande are important boundaries between the United States and other countries. Along the St. Lawrence, the United States and Canada must solve problems of navigation and water use for generating electricity. The major problem confronting the United States and Mexico in connection with the use of the Rio Grande is the sharing of water for irrigation. The longstanding dispute about the boundary between the two countries due to a shifting of the course of the Rio Grande at El Paso, Texas, ended when the Johnson administration came to an agreement with Mexico.

FIGURE 16–2 *The Rio Grande flows between El Paso, Texas, and Ciudad Juarez, Mexico. This segment of the river was once a controversial border between the two countries because the river had shifted its course. After a settlement during the Johnson administration, the channel was lined with concrete to keep the river from shifting again.*

Now the stream bed in that vicinity is permanently contained in a channel lined by concrete (Fig. 16–2).

Some rivers that cross boundaries are the Colorado in North America and the Rhine and Danube in Europe. The latter traverse heavily industrialized sections of Germany where they are thoroughly polluted before entering other countries. While efforts are being made to improve the quality of the water, it remains discolored with waste.

Large rivers influence the quality of bodies of water into which they empty. Before the Aswan High Dam was built on the Nile, the river regularly flooded its banks and carried silt, nitrates, and phosphates into the eastern Mediterranean. The enrichment caused flourishing blooms of phytoplankton, which in turn supported a luxuriant fauna including sardines. The sardine industry thrived and was an important part of the economy. Upon completion of the dam, the flooding and the resultant euthrophication of the sea were halted. The sardine harvest quickly dropped more than 50 percent, and a further decline is expected that will eventually destroy the industry.

With most countries recognizing territorial limits of three miles offshore, the vast oceanic expanse is free territory for any who want to use it. Since its natural resources are open for exploitation, many countries do so. Among the products sought are fish, shrimp, turtles, whales, and oil. Minerals and other products will become important in the future.

That overkilling of animals is possible in oceans is accentuated by the near extinction of whales, the decline of the harvest in certain fishing waters (like the area between Hudson Canyon and Nova Scotia), and the decrease in number of sea turtles. Sea turtles are especially vulnerable because of their habit of laying eggs on land where both the turtles and their eggs are easily exploited. The green turtle used to breed along the North American continent from Massachusetts southward but is now no longer found there.

The hazard of oil spills and leaks has increased of late. Leaks from offshore wells like those at Santa Barbara in January, 1969, and in the Gulf of Mexico in 1971 are unusually destructive—the Santa Barbara slick covered 800 square miles of oceanic surface. In addition, a number of tankers have capsized or sprung leaks, causing large slicks. Such slicks kill wildlife and algae, and beautiful beaches are littered with smelly, sticky oil and decaying plants and animals. The cleanup following a spill is very expensive in both labor and materials.

When migratory animals cross borders, they become the responsibility of two or more political units. They require protection at breeding sites and along their routes of travel, but this is often lacking. The need for cooperation in protecting valuable wildlife and environmental resources is clear: Mammals are being exterminated at a rate of about one species per year. There is also great merit in joint research, information sharing, and monitoring of the environment. Joint ventures bind people together in a common cause, pool talent, and reduce costs.

Many industrial activities are affecting the atmosphere throughout the world. The input of all of these influences into a single swirling envelope of gas is carried beyond the place of origin and beyond political borders. Pollutants quickly mix in the air and blanket the earth, where they threaten to intercept enough radiant energy to affect world climate. The dispersal of radioactivity from atomic explosions, for example, is so frightening that self-interest has imposed genuine restraint on further expositions that would increase atmospheric contamination.

In addition to pollution, changes at the interface between the atmosphere and the earth's surface may drastically alter temperature or evaporation rates, which in turn affect weather. Such changes include urbanization, removal of vegetation, irrigation, and drainage. If any manipulation of the natural environment causes changes in weather patterns, the results are of concern to the world community. In particular, the consequence of weather making (modification) on atmospheric circulation are so poorly understood that a cautious approach should be taken lest a good result in one place produce a bad one elsewhere. To what extent, if any, does rainmaking in one place deny rain to another locality? Breaking up storms such as hurricanes generally meets with approval because of their destructiveness. Similarly, antifogging procedures around airports are accepted with little or no objection.

To date, no methods have been found to permanently change large-scale weather patterns, nor are any in sight. Yet, despite the risks involved, individuals and groups in various countries experiment with weather modification, especially seeding clouds to make rain. During the 1971 drought in several states of western America, rainmaking was frequently tried. The armed forces gained considerable experience in this technology in Laos, where they seeded clouds to increase the rainfall along the Ho Chi Minh trail in an effort to impede the movement of military hardware from North Vietnam.

Channels of cooperation among different countries have never been more open than they are now. There are many avenues for bilateral cooperation between one nation and another; among them are multinational organizations like the United Nations (UN), North Atlantic Treaty Organization (NATO), International Biological Program (IBP), Economic Commission for Europe (ECE), Organization for Economic Cooperation and Development (OECD), and International Union for the Conservation of Nature and Natural Resources (IUCN). Together they cover a wide range of international activities.

The United Nations operates a number of agencies with wide-ranging responsibilities. These include the Economic Commission of Europe (in part); Intergovernmental Maritime Consultative Organization (IMCO); United Nations Educational, Scientific, and Cultural Organization (UNESCO); World Health Organization (WHO); Food and Agriculture Organization (FAO); World Meteorological Organization (WMO); International Atomic Energy Agency (IAEA); and United Nations Conference on the Human Environment (UNCHE).

International efforts are hampered by lack of authority to take action. Almost all that can be done cooperatively is to hold conferences, study problems, make recommendations, and foster treaties. The endeavors frequently consist more of talk than of action. Nevertheless, the exchange of ideas brings the problems into focus and publicizes them. Some of the efforts mentioned have tended to foster an international cooperative attitude and to minimize national considerations.

EDUCATION

Responsibility for protecting the environment cannot be delegated to the handful of zealous naturalists and conservationists who have assumed leadership in the past. They have done a superb job of alerting the public to the impending destruction of unique natural areas and of finding ways to preserve them for the future. In many cases they have fought thankless battles against representatives of special interests who were more concerned about a fleeting profit than about perpetual usefulness. However, population pressures have quickened the pace of change to the point where this small group of guardians of the natural heritage must have additional support.

The purpose of education in general is to raise the level of life as far as possible beyond mere existence, and the purpose of environmental education in particular is to preserve and appreciate the natural harmony that has been millions of years in the making. Educating for the sake of conservation is a commendable objective, but this should not be the only goal. The educational process will be more successful if it includes both understanding of natural processes and motivation to action.

Environmental education is needed by both the child who does not know and the adult who does not care. Of the two, the child is by far more amenable to correction. Something can be done for the child through the public schools, many of which are already incorporating valuable ecological information into their curricula. Some even operate camps where children can work in natural situations and see firsthand some of the relationships between organisms and their surroundings.

The adult who does not care does most of the damage, is outside the influence of formal education, and often has a closed mind when it comes to reforming old habits. The job of educating children is easier because they are responsive and relatively free from prejudice. The teacher can capitalize on their natural curiosity, provided he or she is knowledgeable about ecology; outside of the classroom, children have easy access to a wealth of good books and other sources of information.

As for adults, they can become informed through newspapers, magazines, television, radio, workshops, billboards, and the like. Unless they are completely out of touch, they could hardly be unaware of the current concern for the ecosystem, now national in scope.

Man has often acted as if he were outside the environment instead of part of it. He must become sensitive to organismic and environmental relationships. He must appreciate what he has often taken for granted—that he is dependent on his surroundings. Man, the air he breathes, the water he drinks, and all the organisms of the food web are one.

A major problem of educating the public is presenting the proper information without generating undue emotionalism. Sometimes people imagine when they see a dead animal or notice leaves dropping from a tree that the environment is coming apart, although the animal could have been killed by a cat and the leaves falling because of dry weather. A recent television program portrayed the case of a town whose residents were upset about pollution caused by a local industry. One of the complainers lifted a branch of an evergreen tree, pointed back toward the center of the tree, and said, "Look under here; see how the factory is killing all of the leaves." For the first time he had noticed the phenomenon of leaf-fall, normal in shade and behind-tip locations, but he attributed it to pollution. A town in Connecticut was thrown into panic by an invasion of gypsy moths that defoliate trees. Some people were ready to spray the countryside without thinking about the shotgun effect on pollinating bees, small pets, and birds. Had they known that healthy trees refoliate in time, they might have been undisturbed.

Whether by necessity or desire, man will have to pay more to improve and maintain the condition of his environment. Accustomed to seeking the most for his money, he must be reeducated to accept additional costs for the products he buys in exchange for a better place to live. This will require many economic readjustments, some of which will be painful.

Who will be the educators? No doubt scientific knowledge will find outlets through many sponsoring agencies, which will emphasize such things as recreation, health, and beauty according to their objectives. Some sponsors will be individuals; others, private or public organizations. Education is a big undertaking that needs the services of almost any well-meaning sponsor.

In addition to using educational outlets like schools, mass media, and government agencies, there are a number of other opportunities to educate the public. The home is a good place to train children, and the home town is the logical place to involve local citizenry in projects. Once involved in correcting problems and making improvements, humans seem to experience satisfaction and to develop loyalties. Local industry should by all means become the teacher instead of the target in environmental matters. Any local museum, park, zoo, or botanical garden would be an ideal focal point for special lectures, demonstrations, or movies. Action groups are good if they are properly motivated and unencumbered by emotionalism.

Educational activities of state or national scope require administration and coordination. Already functioning in this field are the U.S. Office of Education, the Department of Health, Education, and Welfare (HEW), and the National Park Service, among many organizations. Sometimes such organizations act as sponsors by awarding funds, and they may publish and distribute literature. Money has been provided by the U.S. Office of Education and HEW to help the Corporation of Public Broadcasting establish an Environmental Center whose purpose is to provide broadcast materials for radio and television. The National Park Service partially funds a National Environmental Education Development (NEED) program that has produced study materials and teacher guides for the elementary schools. These materials should help develop awareness and understanding of the total environment. The Park Service also has a program called NESA (National Environmental Study Areas) through which it is establishing study centers jointly with local communities. Many other new programs are sure to become operational in the near future.

Private organizations, the backbone of past efforts to preserve the environment, have received additional public support and will expand their activities. Such groups have published materials, sponsored special educational programs, and participated in conferences and workshops. Among them are the National Audubon Society, the Sierra Club, the Izaak Walton League, the American Forestry Association, the National Recreation and Park Association, and the Conservation Foundation. Add to these a number of newcomers, and the amount of private involvement in environmental activity is impressive.

ACTION SUGGESTIONS

Never before in history has so much been said about environmental abuse and what needs to be done about it. But the real test of the sincerity of the resolve to improve the environment will come from personal involvement. A selected list of action suggestions follows.

1. Resolve to leave your part of the world a better place than you found it.
2. Make it your business to learn more about nature and take the time to enjoy your surroundings. Beauty does not have to be exotic or distant.
3. Do not accept without question the contamination of the atmosphere, lakes, streams, or oceans. Remember that some substances may have no measurable immediate or long-term effect on the environment.
4. Place less emphasis on standard of living and more on quality of life.
5. Contribute to the development of a public conscience against destructiveness, waste, and disorder (pollution, litter, etc.).
6. Question present attitudes toward land usage. Sprawling building complexes and highways are wasting a valuable resource, the land surface.
7. Question the environmental impact of every proposed land development. Is the project really needed or is it simply a scheme to enrich the developer?
8. Give first consideration to using biological organisms for pest control (e.g., insect control) and recycling (e.g., sewage treatment) before turning to chemical or industrial means.
9. Avoid purchasing products that cannot be readily recycled.
10. Seek ways to reduce needless consumption of both energy and material goods.
11. Strive for firm, uniform application of quality standards by governmental bodies.
12. Encourage your community to plan carefully for its economic and environmental needs. This includes the establishment of

rigorous zoning restrictions and the determination of its optimum population size and density.

13. Insist that fundamental ecological instruction be incorporated into school curricula.
14. Support planned parenthood as a means of achieving personal happiness and responsible parenthood.

REFERENCES AND FURTHER READINGS

Battan, L. J. 1969. *Harvesting the Clouds.* Anchor Books, Doubleday, Garden City, N.Y.

Commoner, B. 1967. *Science and Survival.* Viking, New York.

Darling, F., and J. P. Milton, eds. 1966. *Future Environments of North America.* Natural History Press, Garden City, N.Y.

Handler, P., ed. 1970. *Biology and the Future of Man.* Oxford University Press, New York.

Johnson, H. D. 1970. *No Deposit–No Return; Man and His Environment: A View Toward Survival.* Addison-Wesley, Reading, Mass.

Kohlmeier, L. M., Jr., 1969. *The Regulators, Watchdog Agencies, and the Public Interest.* Harper & Row, New York.

Love, G. A. and Rhoda M. Love, eds. 1970. *Ecological Crisis.* Harcourt Brace Jovanovich, New York.

McHarg, I. L. 1969. *Design with Nature.* Natural History Press, Garden City, New York.

Nash, R. 1967. *Wilderness and the American Mind.* Yale University Press, New Haven, Conn.

Shepard, P. 1967. *Man in the Landscape.* Random House, New York.

Wagner, R. H. 1971. *Environment and Man.* Norton, New York.

Glossary

abiotic nonliving components of the environment
anaerobic organisms living in the absence of free oxygen; respiration in the absence of free oxygen
aphotic stratum (zone) all water below the level to which light penetrates
autecology a branch of ecology dealing with the study of individuals
autotrophic producing its own food; photosynthetic plants
autotrophs plants that synthesize their own food by either photosynthesis or chemosynthesis
background radiation natural radioactivity originating in space and in various earth substances
bald a treeless area on high slopes and peaks in a region where the natural vegetation is a forest; covered with shrubs or herbaceous plants
banks large elevated areas of the ocean floor, often ideal fishing sites
benthic on or in the bottom of a body of water
benthos aquatic organisms on or in the bottom; also the bottom of a body of water
biocides poisons like herbicides, fungicides, or insecticides that kill living organisms
biomass the quantity of living matter, usually of a specific area
biome the largest clearly identifiable community occupying a climatic region and characterized by organisms with a specific life form
biosphere all the life on the earth
biota collectively, the plants and animals of a region
biotic pertaining to life; biological
biotic potential the capacity for a population of organisms to increase in number in a given period of time
bloom the rapid increase of a population such as an alga in a lake
bog an acid marsh, often vegetated with sphagnum and heaths
canopy ordinarily, trees forming the upper stratum in a forest; the overlapping crowns of the tallest trees in a forest

carnivore an animal that eats other animals; a meat-eating animal

carrying capacity the quantity of certain types of organisms that a particular location will support

catabolism metabolism that breaks down complex molecules into simpler ones

chemosynthesis synthesis of foods by using the energy of chemical reactions, as in some types of bacteria

circadian rhythm cyclic behavior approximately twenty-four hours in duration

climax the steady-state community of a successional series that is highly stable and self-perpetuating as long as climatic conditions remain the same

clone the vegetatively (asexually) produced offspring of a single individual

community a group of different kinds of organisms living together in the same place

compensation point (level) in plants, the light intensity allowing the rate of photosynthesis to equal the rate of respiration; under a specific intensity of light, the rate of oxygen consumption from respiration is equal to the rate of oxygen release from photosynthesis

competition interference among populations or individuals of the same population for the limiting factors of the environment

composites plants belonging to the large composite family exemplified by asters, daisies, dandelions, and goldenrod; with compact clusters of small flowers that superficially resemble a single flower

composting converting organic waste into a product suitable for fertilizing or conditioning soil

consumer an organism other than green plants and chemosynthetic bacteria that obtains its nourishment from other living organisms, excrement, or dead bodies

continental shelf a continental fringe shallowly covered by the ocean; the bottom of the neritic zone of the ocean

continental slope the floor of the oceanic basin from the edge of the continental shelf to the deep bottom; steep sides of the oceanic basin below the outer edge of the continental shelf

Coriolis effect the deflection of currents of air and water caused by the rotation of the earth and resulting in cyclonic atmospheric and whirlpoollike patterns

curie a unit of radioactivity defined as 3.7×10^{10} disintegrations per second

deciduous shedding, usually of leaves

deciduous forest a type of forest in the middle latitudes characterized by trees that shed their leaves in the winter

decomposer an organism that causes decay, a process essential to chemical recycling

desert an arid region where under the best conditions the plants are widely spaced and the annual rainfall is usually less than ten inches

diastrophism the crustal movements of the earth that produce topographical features like continents, oceanic basins, and mountains
diatomaceous ooze a soft deposit on the ocean bottom consisting of the siliceous remains of unicellular algae known as diatoms
disclimax a climax maintained by continuous or repeated disturbance such as fire, disease, lumbering, or grazing
dominance the state of exerting the strongest influence
dominants of organisms in a community, those that exert the most influence
doubling time of populations, the amount of time elapsing before the population doubles
dystrophic unproductive (infertile) because nutrients are locked in undecayed vegetable matter; used to describe lakes in cold climates where anaerobic respiration prevails
ecological niche the position occupied by a type of organism relative to all other organisms and the physical environment in which it lives
ecology the science dealing with the relationships between organisms and environments
ecosphere all of the organisms of the earth and the physical factors with which they are associated
ecosystem a more or less self-contained community of organisms together with the environment in which it lives
ectothermic of animals, having a body temperature that fluctuates with the temperature of the surroundings; poikilothermic, or cold-blooded
edaphic of or pertaining to the soil
effluent outflow; frequently refers to the wastes carried by air or water from factories or sewage treatment facilities
endemic native to a region
endothermic of animals, having a steady body temperature internally regulated; homoiothermic, or warm-blooded
energy flow the transfer of energy from one organism to another through the food chain
environment all of the external biological and physical conditions that influence organisms
eolian transported by wind
epilimnion the upper warm-water stratum of a deep lake, located above the metalimnion (thermocline)
eutrophic enriched by the addition of nutrients contained in pollutants, especially descriptive of aquatic environments
eutrophication the enrichment of environments, especially aquatic ones; the addition of nutrients contained in pollutants
facultative able to live under different conditions—e.g., parasitic and free-living or aerobically and anaerobically
food chain a sequence of organisms through which energy (food) flows in an ecosystem
food web the interlocking pathways by which energy flows within a community; food relationships of organisms within a community
foraminiferal ooze a soft deposit on the ocean bottom consisting of the calcareous remains of protozoans known as foraminiferans

forbs herbaceous plants other than grasses

formation the largest clearly identifiable community occupying a climatic region and characterized by organisms with a specific life form; equivalent to a biome

gallery forest an extension of the tropical rainforest along the banks of small streams into drier areas; so named from the overlapping tops of trees on each bank that form a gallerylike enclosure over a stream

gene pool all of the different kinds of genes contained in the individuals of a population

genus a group of closely related species; the first of two words in the scientific name of a plant or animal

germ-plasm bank a collection of variable but interbreeding organisms containing a wide assortment of genes that may be used to produce new varieties in the future

green revolution agricultural developments such as breeding, cultivation, and use of fertilizers that have drastically increased crop yields

guyots truncated (flat-topped) volcanic cones arising from the floor of the ocean but below the surface of the water

habitat the place where an organism lives; the kind of place where an organism is ordinarily found

half-life the time required for half of the atoms of a radioactive element to disintegrate

heaths plants of the heath family, including such well-known shrubs as azaleas, rhododendrons, cranberries, and blueberries

herbicides poisons that kill plants; weed killers such as 2,4-D

herbivore an animal that eats plants

heterotrophic dependent on other organisms for nourishment

heterotrophs organisms that depend on other organisms for their food; all animals and some plants that are unable to synthesize their own foods

homeostasis the maintenance of a steady-state condition

hydrosphere the aqueous part of the earth as distinguished from the atmosphere and the lithosphere

hypolimnion the lower cold-water stratum of a deep lake, located below the metalimnion (thermocline)

incaparina a protein concentrate with a high cottonseed and cornmeal content

intertidal zone the coastal area flooded at high tide and exposed at low tide; the strip of land between the high- and low-tide marks

ionizing radiation radiations that cause the formation of ions (electrically charged particles)

isotope one of the variant atoms of an element, differing in the number of neutrons in its nucleus

J curve a population growth curve shaped like a J and depicting a rapid increase in a population, characteristic of populations with high biotic potential

Karst region any location having soluble rock strata such as limestone and characterized by caves and sinkholes

lake a large inland body of open water

laubina a protein concentrate with a high wheat and chick pea content
legumes plants belonging to the pea family, exemplified by beans, clover, alfalfa, and locust
lentic with standing water; used in connection with freshwater habitats
life form distinctive type of organisms characteristic of a specific environment—e.g., grasses of prairies
limnetic in or of fresh water
limnetic stratum (zone) in lakes, all open water above the bottom; according to some authors, an upper stratum of a lake above the profundal stratum
littoral zone the shallow-water marginal zone of a lake where rooted vegetation often grows; the intertidal zone of the ocean
loess fine-grained deposit of windblown dust
lotic with flowing water; used in connection with freshwater habitats
mangrove forest a tropical forest of the intertidal zone consisting of entangled trees with stilt roots and broad evergreen leaves
mariculture growing marine organisms where the habitat is artifically improved
marsh a place with shallow standing water or waterlogged soil, vegetated but treeless
Mediterranean-type scrub forest a type of forest characterized by evergreen trees and shrubs with small, thick leaves living in temperate zones where the growing season is in the winter and the dormant (dry) season is in the summer
metabolites compounds produced by metabolism
metalimnion the thermocline of a deep lake; the zone of rapid vertical change in temperature located between the epilimnion and the hypolimnion of a deep lake
microcosm a small ecosystem
monoculture the presence of a single kind of organism in a particular area; man frequently develops monocultures in agriculture and tree farming
mortality death rate
muskeg a bog in the northern coniferous forest
natality birth rate
neap tide the lowest tide, occurring when the gravitational fields of the moon and sun counteract each other because of their positions with respect to the earth
nekton free-swimming organisms such as fishes
neritic zone the part of the ocean above the continental shelves
neuston aquatic organisms that live on or suspended from the surface film of water
niche see **ecological niche**
northern coniferous forest a type of forest in the Northern Hemisphere characterized by cone-shaped evergreen trees with needlelike leaves, located in a belt south of the tundra and north of the deciduous forest
obligate limited to a single way of life—e.g., as a parasite or under anaerobic conditions

oligotrophic impoverished; descriptive of deep lakes where green plants are relatively scarce

ooze a soft deposit on the ocean floor

pedon aquatic organisms that live on or in the bottom; generally used in connection with lake habitats

pelagic oceanic; in or on the open ocean

periphyton aquatic organisms that live attached to plants or other objects; generally used in connection with lake habitats

permafrost permanently frozen subsoil of the tundra

pH a scale indicating the degree of acidity or alkalinity of a solution; the scale ranges from 0 to 14 with 7 indicating neutrality, above 7 alkalinity, and below 7 acidity

pheromone a substance secreted into the environment by an individual and perceived by other individuals of the same species, who react to it in their behavior or development

photic stratum the upper layer of water penetrated by light

photoperiod the relative amount of exposure to light and darkness that triggers or controls various biological processes

photosynthesis the synthesis of food by green plants in the presence of light

physiological drought a condition in organisms, particularly plants, when water is not available to tissues in spite of external abundance—e.g., when the water of soil is frozen and plants are unable to obtain it to meet losses resulting from transpiration

phytoplankton minute aquatic plants that float on or near the surface

pioneers the first organisms to inhabit an area

plankton small aquatic organisms, mature or immature, living near the surface and wafted about by water currents

playa the bottom of a desert basin; an intermittent lake basin often encrusted with salty deposits

pollution any abnormal environmental condition that upsets the natural equilibrium

pond a small body of open water, often artificial and often having rooted plants over much of its bottom

population all individuals of a species or all members of the same species in a specific location

population density the number of individuals of a population in a specific location

population explosion a rapid increase in population

population implosion the concentration of humans in small areas

population pressure the demands made by a population on its environment or on the members of the same population

prairie a middle-latitude grassland where the grass is characteristically tall

predator an animal that preys on or consumes other animals

primary consumers animals that use green plants for food—e.g., herbivores

primary productivity the rate of increase in biomass as a result of photo-

synthetic or chemosynthetic activity; the amount of biomass produced by autotrophic organisms in a specific period
primary treatment in sewage disposal, the initial step, in which some of the solids are removed by settling
procaryotic organisms like bacteria and blue–green algae that do not have a well-defined nucleus
producer any organism that synthesizes or assimilates foods and increases in size or number
production the amount of accumulated organic matter
productivity the rate of increase in biomass of an individual, a population, or a community of organisms
profundal zone the bottom of a lake, beneath the hypolimnion; according to some authors, the zone also includes the water above the bottom
pteropodial ooze a soft deposit on the ocean bottom consisting of the calcareous remains of a group of gastropod mollusks known as pteropods
pyramid a cone-shaped graph showing the quantity (energy, biomass, or number) produced at different trophic levels
rem Roentgen Equivalent in Man, or the amount of ionizing radiation necessary to produce the same biological effect as one roentgen of high-penetration X rays
roentgen a unit or quantity of X or gamma radiation; the amount of X or gamma radiation that produces ions in one cubic centimeter of air carrying one electrostatic unit of charge
salpingectomy the surgical removal of oviducts or portions of them; sterilization of the female by this process
salt pan the salty encrustation remaining after all water evaporates from a tidal pool
saprophage an organism that feeds on dead plants and animals
savanna a tropical grassland
scrub forest a tropical forest of stunted, frequently thorny trees, often scattered, growing where the rainfall is barely enough to support them
S curve a population growth curve shaped like a slanting S and characteristic of population increase following introduction of a species into a new habitat or the resumption of favorable environmental conditions
Secchi disc a circular plate lowered into water to measure turbidity
secondary consumers animals that eat primary consumers; in the food chain, the second link beyond the producers
secondary productivity the rate of increase in biomass of consumers and decomposers
secondary treatment in sewage disposal, the step in which bacteria break down organic contaminants; to facilitate the decomposition, oxygen is usually added by agitating the water
seiche shore-to-shore oscillation of water in a lake, usually initiated by wind
semideciduous forest a tropical forest growing where there are distinct wet and dry seasons; the broad-leaved trees may lose many or all of their leaves during the dry season

sinks sinkholes; natural depressions in limestone regions resulting from solution of rock or collapse of a cave roof

species a population of similar individuals, usually capable of reproducing freely among themselves

spring tide the highest tide, occurring when the moon, earth, and sun are in line and the gravitational fields of the moon and sun act together

standing crop the biomass of a specific place and time

steppe a middle-latitude grassland where short grasses predominate

subclimax a prolonged successional stage preceding the climax community; sometimes used for any successional stage preceding the climax

sublittoral zone the part of the bottom of a body of water that lies below the littoral zone; as used here, the bottom of a lake from the zone of rooted vegetation to where the hypolimnion intersects the bottom

succession the replacement of one community by another

swamp a place with shallow standing water or waterlogged soil, containing trees

symbiosis in a broad sense, different kinds of organisms living in a close relationship; in a narrow sense, different kinds of organisms living in a close relationship from which all members benefit (mutualism)

synecology a branch of ecology dealing with the study of communities

taxon any taxonomic category such as a phylum or a species

temperate rainforest a middle-latitude forest characterized primarily by evergreen conifers and supported by heavy precipitation consisting of cool rain or snow

temperature inversion in water, a colder stratum overlying a warmer one; in air, a colder stratum trapped beneath a warmer one

territoriality a behavior characteristic of some animals in which specific areas are claimed and defended for their own use

tertiary consumers animals that eat secondary consumers; in the food chain, animals that are three links beyond the producer stage

tertiary treatment in sewage disposal, the final and most expensive step, in which the hard-to-remove contaminants are extracted by special means

thermal enrichment unnatural increase in the temperature of an environment (when considered beneficial)

thermal pollution unnatural increase in the temperature of an environment (when considered harmful); usually used in connection with water environments heated by an atomic reactor or an industrial process

thermocline the metalimnion of a deep lake; the zone of rapid vertical change in temperature located between the epilimnion and the hypolimnion of a deep lake

thermophiles organisms tolerant of the high temperatures of hot springs and streams; organisms that live in hot springs and streams

tolerance level the limits of the range of conditions in which an organism can survive

trophic level a nutritional level—e.g., green plants (primary producers), herbivores, or carnivores

trophogenic stratum the upper part of a lake, where photosynthesis is the predominant metabolic activity and building up exceeds tearing down

tropholytic stratum the lower part of a lake, where respiration is the predominant metabolic activity and tearing down exceeds building up

tropical rainforest a lush tropical forest receiving year-round rainfall and characterized by trees with broad evergreen leaves

troposphere the portion of the atmosphere between the earth and the stratosphere

tundra a cold region of the far north where the subsoil is permanently frozen and the characteristic plants grow in ground-hugging mats; tundralike zones above the timberline of high mountains (alpine tundra); the dwarfed vegetation characteristic of extremely cold climates

turbidity cloudiness of water due to light-blocking impurities or small organisms

turnover vertical circulation in bodies of water, especially lakes

understory stratum of vegetation growing to a height just under that of a forest canopy

upwelling the rising of cold water from the depths to the surface in some parts of the ocean, frequently along the margins of continents

vasectomy the surgical removal of portions of the sperm ducts; sterilization of the male by this process

vector the transmitter of a parasite

xerophytic plants adapted to arid conditions

yield the amount harvested; the portion of biomass that is usable

zooplankton minute aquatic animals, mature or immature, that live near the surface and are wafted about by water currents

Appendix 1.
Habitats, Distinctive Organisms, and Communities

Through the years all states in the nation have reserved natural areas for the benefit of their citizens. Some of them have unique biological or geological features. Others are typical countryside tracts of land made available to the public. The acquisition of these valuable resources was the result of the wisdom of many farsighted people who observed the rapid and widespread alteration of nature and wanted to preserve some of its beauty for the pleasure of future generations. In addition, many of these natural habitats were set aside as wilderness areas for the protection of wildlife from man. Because of a growing public awareness of the importance of such areas, these early efforts to preserve land have been accelerated. Ecologists, who recognize the delicate interrelationships among all components in a nature-regulated ecosystem of which mankind is a part, endorse the movement.

Appendix 1 is a list of natural areas located in all of the fifty states. It is extensive but, because of space limitation and new acquisitions, not complete. It lists many parks, forests, preserves, refuges, natural areas, wild areas, and recreational areas where examples of types of organisms and habitats of ecological interest can be found. The list is arranged by state for the convenience of local residents who want to locate sites near their homes or for use by travelers who may be passing through a particular area. In each case, distinctive communities and habitats are indicated, as well as the presence of nature trails and availability of information for visitors.

Name of park or preserve	Distinctive habitats or communities	Nature trails	Information available to guide visitors
ALABAMA			X
Cheaha State Park	rhododendron, flame azalea, and oak-leaf hydrangea deer, small game	X	
Chewacla State Park	located on the Fall Line separating the Piedmont Region from the Coastal Plains many native plants	X	X
DeSoto State Park	Little River Canyon, deepest east of Mississippi River, falls rhododendron, mountain laurel	X	X
Gulf State Park	sandy beaches, freshwater lake many unusual forms of plant life	X	X
Joe Wheeler State Park	variety of plant life fishes, small game	X	X
Lake Guntersville State Park	variety of plant life deer, small game	X	X
Monte Sano State Park	unusual mountain formation distinctive plant life	X	X
ALASKA			
Battery Point Recreation Area	rocky and sandy beaches coniferous forest, muskeg bald eagles		

Chugach State Park	wilderness areas, mountain zonation, alpine tundra	X	
	moose, Dall sheep		
Denali State Park	braided stream		
	coniferous forest, muskeg		
	moose, black bear		
Kachemak State Park	wilderness, glaciers, fault zones		
	alpine tundra, mountain zonation, coniferous forests		
	moose		
Katmai National Monument	volcanic region		
Kenai National Moose Range	moose		
Mt. McKinley National Park	tundra		
	caribou, willow ptarmigan		
ARIZONA			
Bankhead National Forest	virgin hardwood forest		
Canyon de Chelly National Monument	sand dunes	X	
Grand Canyon National Park	youthful stream	X	X
	ponderosa pine		
Joshua Tree Forest	Mojave Desert		
Kaibab National Forest	ponderosa pine, pinyon pine, juniper		
Kitt Peak	oak forest		
Kofa Mountains	Sonoran Desert		
	bighorn sheep		
Mazatzal Wilderness Area	chaparral		
Mt. Baldy Wilderness Area	mountain zonation, coniferous forest, alpine tundra		
Oak Creek Canyon	youthful stream		

Name of park or preserve	Distinctive habitats or communities	Nature trails	Information available to guide visitors
Oak Mountain State Park	azalea, mountain laurel, and many other flowering plants		
	deer, turkey, and other small game	X	X
Organ Pipe National Monument	Quitobaquito desert oasis		X
	mesquite, cactus		
Saguaro National Monument	cactus	X	X
San Francisco Mountains	mountain zonation, pinyon pine, fir, spruce		
Santa Catalina Mountains	mountain zonation		
Superstition Wilderness Area	Sonoran Desert		
Willcox Dry Lake	salina		
ARKANSAS			
Buffalo River State Park	caves, wild river	X	X
	oak–hickory forest		
Devil's Den State Park	caves, wilderness area, mature stream	X	X
Petit Jean State Park	caves, wilderness areas	X	X
	oak–hickory forest		
Queen Wilhelmina Park	mountain forest	X	X
Village Wilhelmina Park	wilderness area		
	virgin vegetation		
CALIFORNIA			
Anza–Borrego Desert State Park	Joshua tree woodland, palm oasis	X	X
Ano Nuevo State Reserve	rocky littoral		

Armstrong Redwoods State Reserve	coastal redwood forest	X	X
Asilomar State Beach	coastal strand	X	X
Azalea State Reserve	coastal redwood forest	X	X
Big Basin Redwoods State Park	coastal redwood forest	X	X
Bothe–Napa Valley State Park	coastal redwood forest, California chaparral, northern oak woodland		
Calaveras Big Trees State Park	sequoia forest	X	X
Death Valley National Monument	sand dunes, salt flats	X	X
Del Norte Coast Redwoods State Park	coastal redwood forest	X	X
Gaviota State Park	southern oak woodland, California coastal chaparral	X	X
Humboldt Redwoods State Park	coastal redwood forest	X	X
Huntington State Beach	sandy beach least tern sanctuary		
Inyo National Forest	bristlecone pines, the oldest living organisms	X	X
Jedediah Smith Redwoods State	coastal redwood forest	X	X
John Little State Reserve	rocky littoral and sublittoral		
Joshua Trees State Park	Joshua tree woodland	X	X
Kruse Rhododendron State Reserve	mixed evergreen forest	X	
Leo Carillo State Beach Reserve	sandy beach		
McArthur–Burney Falls Memorial State Park	youthful stream	X	X
MacKerricher State Park	sandy and rocky beach north coastal prairie	X	X
Mailliard Redwoods State Reserve	coastal redwood forest	X	X
Malakoff Diggins State Historic Park	montane chaparral	X	X
Malibu Lagoon State Beach	estuarine littoral and sublittoral		
Marin Headlands State Park	northern coastal scrub		

Name of park or preserve	Distinctive habitats or communities	Nature trails	Information available to guide visitors
Marshall Gold Discovery State Historical Park	northern oak woodland	X	X
Montana de Oro State Park	California coastal chaparral, coastal sage, scrub	X	X
Montgomery Woods State Reserve	coastal redwood forest	X	X
Morro Bay State Beach	saltwater marsh	X	X
	coastal sage scrub		
Morro Strand State Beach	coastal strand (dunes)		
Mount Diablo State Park	north oak woodland		X
Mount San Jacinto Wilderness State Park	California chaparral, ponderosa pine forest, alpine fell	X	X
Mount Tamalpais State Park	central valley prairie	X	X
Palomar Mountain State Park	ponderosa pine forest, southern oak woodland	X	X
Patrick's Point State Park	rocky littoral	X	X
	kelp forest		
Pfeiffer Big Sur State Park	coastal redwood forest	X	X
Pismo State Beach	coastal strand, freshwater lagoon, marsh, swamp	X	X
Plumas–Eureka State Park	red fir forest	X	X
Point Lobos State Reserve	rocky littoral and sublittoral	X	X
	kelp forest, Monterey and Gowen cypress forests, Monterey pine forest		
Point Mugu State Park	Central Valley prairie, coastal sage scrub, California coastal chaparral	X	X

Point Sal State Beach	rocky bluffs, sandy beach		
	coastal sage scrub		
Prairie Creek Redwoods State Park	coastal prairie, coastal redwood forest	X	X
Providence Mountains State Recreation Area (Mitchell Caverns State Reserve)	caves	X	X
Richardson Grove State Park	coastal redwood forest	X	X
Russian Gulch State Park	sandy beach	X	X
	pygmy forest		
Salt Point State Park	rocky littoral		
	kelp forest, north coastal prairie		
Samuel P. Taylor State Park	old streams	X	X
Smithe Redwoods State Reserve	coastal redwood forest	X	X
Squaw Valley State Recreation Area	red fir forest, subalpine forest, lodgepole pine forest		
Standish–Hickey State Recreation Area	coastal redwood forest		
Sugarloaf Ridge State Park	California chaparral		
Sugar Pine Point State Park	oligotrophic lake	X	X
	ponderosa pine forest		
Tahoe State Recreation Area	oligotrophic lake, alder swamp	X	X
Tomales Bay State Park	Bishop pine littoral and sublittoral bay communities		
Tule Elk State Reserve	wild animal preserve		X
Van Damme State Park	pygmy forest, fern canyon	X	X
COLORADO			
Bonny Reservoir State Recreation Area	prairie, floodplain (cottonwood stands and marshes)	X	X
	bison, valuable waterfowl area		

Name of park or preserve	Distinctive habitats or communities	Nature trails	Information available to guide visitors
Colorado State Forest	youthful and mature streams, sand dunes, glacial cirque lakes, wilderness areas		
	sagebrush, montane and subalpine coniferous forest, fescue meadows, sloughs, alpine tundra		
	bighorn sheep		
Flat Tops Primitive Area	wilderness areas, wild streams	X	X
Golden Gate Canyon State Park	youthful streams	X	X
	montane coniferous forest, ponderosa and lodgepole pine, Douglas fir, aspen		
Gore Range Eagles Nest	wilderness areas, wild streams	X	X
Great Sand Dunes National Monument	sand dunes		
Gunnison National Monument	youthful streams		
Highline Reservoir	desert shrubland (saltbush, rabbit brush, greasewood, and tamarisk)	X	X
Horse Tooth State Recreation Area	foothills grassland, montane coniferous forest	X	X
La Garita Wilderness Area	wilderness areas, wild streams	X	X
Lathrop State Park	pinyon–juniper woodland, prairie	X	X
	bison		

Maroon Bells–Snowmass Wilderness Area	wilderness areas, wild streams	X	X
Mesa Verde National Park	junipers		
Mt. Zirkel Wilderness Area	youthful streams, oligotrophic lakes	X	X
Rawah Wilderness Area	coniferous forest, virgin vegetation	X	X
Rocky Mountain National Park	mountain zonation, alpine tundra	X	X
San Juan–Upper Rio Grande	wilderness areas, wild streams	X	X
Steamboat Lake State Park (man-made fishing lake)	subalpine coniferous forest (lodgepole pine–aspen), sagebrush flats	X	X
West Elk Wilderness Area	wilderness areas, wild streams	X	X
Wilson Mountains	wilderness areas, wild streams	X	X
CONNECTICUT			
Gillette Castle State Park	mixed hardwood		
Hammonasset State Park	salt marsh, sandy beach		
Penwood State Forest	mixed hardwood	X	
DELAWARE			
Appoquinimink Wildlife Area	freshwater marsh		
Brandywine Creek State Park	piedmont deciduous forest	X	
Cape Henlopen State Park	beach, dunes		
Killen's Pond	freshwater lake		
Little Creek Wildlife Area	salt marsh	X	
Pea Patch Island	heron rookery		
Petersburg Wildlife Area	mixed forest	X	
Trap Pond State Park	cypress swamp	X	
FLORIDA			
Blackwater River State Forest	longleaf pine forest		X
Collier Seminole State Park	wilderness areas	X	X
	cypress swamp, mangrove, virgin vegetation		

Name of park or preserve	Distinctive habitats or communities	Nature trails	Information available to guide visitors
Corkscrew Swamp Sanctuary	virgin cypress swamp	X	X
Everglades National Park	savanna, hammocks, mangrove forest	X	X
Florida Caverns State Park	cave	X	X
	oak–hickory forest		
Highlands Hammock State Park	cypress swamp, prairie (savanna)	X	X
	wild animal preserve, sandhill crane		
Hugh Taylor Birch State Park	sandy beaches	X	X
John Pennykamp State Park	living coral reef		X
Jonathan Dickinson State Park	coastal dunes	X	X
Little Talbot Island State Preserve	sandy beaches, dunes, salt marsh	X	X
Long Pine Key Wildlife Refuge	deer refuge	X	X
Myakka River State Park	mature streams, wilderness	X	X
	wild animal preserve, Florida panther		
Oleno State Park	oak–hickory forest	X	X
Paynes Prairie State Preserve	Florida wild rat, glossy ibis		
St. Joseph State Park	sandy beaches, dunes		X
Torreya State Park	mature stream	X	X
	oak–hickory forest, coniferous forest, Torreya and Florida yew trees		
Waccassassa Bay State Preserve	salt marsh, wilderness area coniferous forest, cypress swamp		

GEORGIA			
Amicalola Falls State Park	oak–hickory forest	X	X
Black Rock Mountain State Park			
Blackburn State Park			
Cloudland Canyon State Park			
Fort Mountain State Park			
Tugaloo State Park			
Vogel State Park			
Crooked River State Park	streams cypress swamp	X	
F. D. Roosevelt State Park	coniferous forest, oak–hickory forest	X	
Bobby Brown State Park			
Mistletoe State Park			
Victoria Bryant State Park			
Hard Labor Creek State Park	streams coniferous forest	X	
High Falls State Park	streams coniferous and hardwood forests	X	X
Indian Springs State Park	sulfur springs, streams oak–hickory forests	X	X
Magnolia Springs State Park	freshwater springs cypress swamp, coniferous and oak–hickory forests	X	X
Okefenokee National Wildlife Preserve	cypress swamps, hammocks alligator	X	X
Reed Bingham State Park	cypress swamp, coniferous forest	X	X

Name of park or preserve	Distinctive habitats or communities	Nature trails	Information available to guide visitors
Stephen C. Foster State Park	wilderness area		
	cypress swamp, coniferous forest, virgin forest		
HAWAII			
Akaka Falls State Park	youthful streams		
Diamond Head State Monument	chaparral		X
Haena State Park	youthful streams, wilderness, sandy and rocky beaches		
Hapuna Beach State Recreation Area	sandy and rocky beaches		
	chaparral		
Iao Valley State Park	youthful streams		
Kahana Valley State Park	youthful and mature streams		
Keaiwa Heiau State Recreation Area	coniferous forest		
Kilauea State Recreation Area	virgin vegetation		
Kokee State Park	youthful and mature streams, wilderness	X	X
	virgin vegetation		
Launiupoko State Wayside	rocky beach		
Makua–Kaena Point State Park	cave, beaches, wilderness		
	virgin vegetation, chaparral		
Mauna Kea State Park	desert		
	savanna, mountain zonation, virgin vegetation		

Na Pali Coast State Park	youthful streams, sandy and rocky beaches, wilderness, chaparral		
Poli Poli State Recreation Area	coniferous forest		
Polihale State Park	sandy and rocky beaches, dunes		
	chaparral		
Pua Ualakaa State Wayside	chaparral		
Waahila Ridge State Recreation Area	coniferous forest		
Wahikuli State Wayside	sandy and rocky beaches		
Waianapanapa State Park	cave, sandy and rocky beaches		
Wailua River State Park	cave, mature streams, sandy and rocky beaches		
Wailuku River State Park	youthful stream, rocky beach (freshwater)		
Waimanalo Bay State Recreation Area	sandy beaches, dunes		
Waimea Canyon State Park	distinctive types of vegetation	X	X
IDAHO			
Bruneau Sand Dunes State Park	inland sand dunes	X	X
	alkali, marsh, and desert vegetation		
Heyburn State Park	whitetail deer	X	X
Idaho Primitive Area (National Forest)	high wilderness, wild river		
Minnetonka Cave (National Forest)	limestone cave	X	X
Salmon River Breaks Primitive National Forest	wild river		
Sawtooth Primitive Area (National Forest)	high wilderness		
	mountain zonation		
ILLINOIS			
Beall Woods Nature Preserve	virgin forest of Wabash	X	X
Ferne Clyffe State Park	beech–maple woods	X	X

Name of park or preserve	Distinctive habitats or communities	Nature trails	Information available to guide visitors
Goose Lake Prairie Nature Preserve	ponds, marsh mesic prairie		X
Heron Pond Nature Preserve	old streams cypress swamps	X	
Illinois Beach Nature Preserve	sandy beaches, dunes prairie	X	X
Pere Marquette State Park	prairie	X	X
Sand Prairie–Scrub Oak Nature Preserve	sand dunes prairie, oak–hickory forest		
INDIANA			
Beaver Lake	wet prairie wild animal preserve		
Beechwood	youthful streams beach–maple and swamp forests	X	X
Bendix Woods	moraines	X	X
Edna W. Spurgeon Woods	beech–maple forest		
Dogwood	beech–maple forest	X	X
Scout Ridge			
Donaldson's Woods	beech–maple and oak–hickory forests, virgin vegetation	X	X
Eunice H. Bryan Preserve	beech–maple, oak–hickory, and swamp forests, virgin forest	X	X
Grider's Woods	oak–hickory forest	X	X

Indiana Dunes State Park	sandy beaches, dunes, freshwater marshes	X	X
	beech–maple, oak–hickory, coniferous and swamp forests, virgin vegetation		
Laughery Bluff	cave, old streams		
	beech–maple forest		
Moraine Preserve	eutrophic lake, kettle holes, moraines		
	beech–maple and oak–hickory forests		
Ogle Hollow	beech–maple and oak-hickory forests	X	X
Pedestal Rock	old streams		X
Pine Hills	beech–maple, oak–hickory, and coniferous forests		
Post Oak–Cedar	hill prairie, oak–hickory forest	X	X
Rocky Hollow–Fall Canyon	old streams	X	X
	beech–maple, oak–hickory, and coniferous forests, virgin vegetation		
Sandhill	oak–hickory and swamp forest		
Tamarack Bog	freshwater marshes, mature streams		
	sphagnum bogs, swamp forest		
Wolf Cave	cave, youthful streams	X	X
	beech–maple and oak–hickory forests		
IOWA			
Bluffton Fir Stand Preserve	balsam fir		
Cayler Prairie	virgin prairie		
Hayden Prairie Preserve			
Kalson Prairie Preserve			
Merritt Forest Preserve			
Sheeder Prairie Preserve			

Name of park or preserve	Distinctive habitats or communities	Nature trails	Information available to guide visitors
Gitchie Manitou	prairie, brushland		
Maquoketa Caves State Park	caves		
Pilot Knob State Park	sphagnum bog with rare plants		
White Pine Hollow Preserve	largest stand of native white pine in state		
Wittrock Indian Village Preserve	prairie		
Woodman Hollow Preserve	rare plants		
KANSAS			
Breidenthal Biological Reserve	oak–hickory forest		
Cedar Bluffs Reservoir	bluestem–grama prairie		
Cimarron National Grasslands	short-grass prairie, sandsage–bluestem on sandhills	X	X
Haskell Wetlands	low wet prairie		
Konza Prairie Research Natural Area	bluestem prairie		
Ross Natural History Reservation			
Quivira National Wildlife Refuge	bald eagle	X	X
Sand Prairie Natural History Reservation	sandhills vegetation		
State Game Preserve, Garden City	bison	X	X
Sterling, Kansas, Salt Marshes	salt marsh		
University of Kansas Natural History Reservation	oak–hickory forest, bluestem prairie		
KENTUCKY			
Carter Caves State Park	caves	X	X
	sphagnum bog, northern relic plants (mixed mesophytic forest)		

Cumberland Falls State Park	northern relic plants (mixed mesophytic)	X	X
General Butler State Park	oak–hickory forest	X	
Pennyrile Forest State Park			
Rough River Dam State Park			
John James Audubon State Park	nature center	X	
Lake Barkley State Park	western mesophytic forest and many southern species		X
Lilley Cornett Woods State Park	wilderness area		
	virgin vegetation		
Mammoth Cave National Park	cave formations	X	X
Natural Bridge State Park	geological features (natural bridges)	X	X
	northern relic plants (mixed mesophytic forest)		
Pine Mountain State Park	hemlock coves, mountain laurel, northern relic plants	X	X
Reel Foot Lake	cypress swamp		
	national wildlife refuge		
LOUISIANA			
Audubon State Park	coniferous and virgin forest	X	X
Chicot State Park	mature streams, wilderness area	X	X
	oak–hickory forest		
Fontainebleau State Park	coniferous forest	X	X
Marksville Prehistoric Indian Park	cypress swamps	X	X
Sam Houston State Park	wilderness areas	X	X
	coniferous forest		
	wild animal preserves		
MAINE			
Acadia National Park	rocky shores, tidal pools	X	X

Name of park or preserve	Distinctive habitats or communities	Nature trails	Information available to guide visitors
Allagash Wilderness Waterway	youthful stream, wild stream, wilderness		X
	coniferous forest		
Baxten State Park	wilderness		X
	mountain zonation, beech–maple forest, alpine tundra, balds		
Camden Hills State Park	rocky shores		
Holbrook Sanctuary			
Two Lights State Park			
Grafton Notch State Park	mountain zonation, beech–maple forest	X	
Mountain Blue State Park			
Moosehorn National Wildlife Refuge	wilderness area	X	X
	beech–maple forest		
Pemaquid Historical Site	natural history		
Popham Beach State Park	sandy beach, dunes		
Scarboro Wilderness Management Area	salt marshes		
Sebago Lake State Park	sandy beach		
Wolf Neck State Park	rocky shore	X	
	old-growth coniferous and beech–maple forest		
MARYLAND			
Assateague State Park	sandy beaches, dunes, salt marshes		X
Cransville Swamp	alpine tundra		X
Deep Creek Lake State Park	eutrophic lake		X

Herrington Manor State Park	old stream	X	X
	oak–hickory and coniferous forest		
Milburn Landing State Park	cypress swamp, loblolly pine	X	X
Potomac State Forest	old stream, wilderness		X
Shad Landing State Park	cypress swamp and distinctive vegetation		X
Swallow Falls State Park	wild streams, wilderness	X	X
	virgin vegetation		
MASSACHUSETTS			
Acushnet Cedar Swamp	virgin Atlantic white cedar		X
Bash Bish Falls State Forest	high waterfall and stream activity		
Bradley Palmer State Park	unique exotic vegetation		
Cape Cod National Seashore	sandy beaches, dunes		
Horseneck Beach State Reservation			
Catamount State Forest	wilderness area		
Chesterfield (Dead Branch)	dystrophic lake		
Connecticut River	old stream		
Harold O. Cook State Forest	Norway spruce plantations		
Lake Quinsigamond State Park	eutrophic lake		X
Mohawk Trail State Forest	coniferous forest, white pines		X
Moore State Park	wetland		
	unique vegetation		
Mt. Greylock State Reservation	beech–maple forest	X	
	Bicknell thrush		
Myles Standish State Forest	oligotrophic lake		
Parker River Area	salt marsh	X	
Pittsfield State Forest	azalea field		

Name of park or preserve	Distinctive habitats or communities	Nature trails	Information available to guide visitors
Plum Island State Park	sand dunes	X	
Rock Port Area	rocky beaches		
Willowdale Forest, Upton Forest	oak–hickory forest		
Windsor State Forest	young streams		
MICHIGAN			
Algonac State Park	nature study area	X	
	prairie		
Besser Natural Area	virgin pine forest		
Interlochen State Park			
Betsy Lake Area	wooded dunes, muskeg, evergreen swamps, open marshland		
Bois Blanc Island	nature study area	X	X
	virgin pine		
Isle Royale National Park	wilderness area	X	X
	moose, wolf		
Laughing Whitefish Falls Scenic Site	youthful stream		
	northern hardwood forest		
Ludington State Park	sandy beach	X	X
P. H. Hoeft State Park	Huron dunes	X	X
	native hardwood trees		
Porcupine Mountain State Park	virgin forest	X	X
Rochester–Utica Recreation Area	shadbush nature area		
Roscommon Red Pine Area Preserve	virgin red pine woodland		

South Manitou Natural Area Preserve	dunes		
Tahquamenon Falls State Park	youthful streams, swamps, sand dunes, moraines	X	
	primeval forest, cedar–white birch complex, muskeg		
Tobico Marsh Preserve	marsh		
Wagner Falls State Scenic Site	youthful stream	X	
	mature northern hardwood climax forest		
Warren Dunes State Park	dunes, nature study area	X	
Waterloo Recreation Area	black spruce–tamarack bog, red maple swamp	X	
Wilderness State Park	sandy beaches, freshwater marsh, wilderness	X	X
MINNESOTA			
Beaver Creek Valley State Park	mature spring-fed streams	X	
Forestville State Park	hardwood and bottom-land forests		
Whitewater State Park			
Blue Mounds State Park	prairie	X	
Boundary Waters Canoe Area	oligotrophic lakes, young streams, rocky shores	X	X
	coniferous forest, virgin vegetation		
Gooseberry Falls State Park	youthful streams, rocky beaches	X	X
	coniferous forest		
Itasca State Park	coniferous forest, maple–basswood forest, virgin vegetation	X	X
Pipestone National Monument	prairie	X	X

Name of park or preserve	Distinctive habitats or communities	Nature trails	Information available to guide visitors
Rice Lake National Wildlife Refuge	wild rice	X	X
	waterfowl		
Sherburn National Wildlife Area	sandy beaches, dunes	X	X
	oak forest		
St. Croix State Park	wild river	X	X
	coniferous and hardwood forests		
Thief Lake Wildlife Area	muskeg, bog forest		
	moose		
MISSISSIPPI			
Bienville National Forest	virgin loblolly pine forest	X	X
	red-cockaded woodpecker		
Delta National Forest	virgin overcup oak, Nuttall oak, red gum stands	X	X
DeSoto National Forest	wild stream	X	X
	coniferous forest		
	salt and brackish marsh	X	
Gulf Islands National Seashore	oak–hickory and shortleaf pine forests		
Holly Springs National Forest	sandy beaches, dunes, salt marsh	X	X
Magnolia State Park	live oak		
	tupelo–cypress swamps	X	X
Malmaison Wildlife Management Area	highlands, loess region	X	
Natchez Trace Parkway	prairie, coastal white cedar swamp, unique vegetation		

National Refuge, Jackson County	coastal terrace		
	savanna, cypress swamp, sphagnum bogs		
	sandhill cranes		
Noxubee National Wildlife Refuges	cypress swamps		
	alligators		
MISSOURI			
Babler State Park	oak–hickory forest	X	
Knob Noster State Park			
Lewis and Clark National Forest			
Big Oak Tree State Park	virgin vegetation	X	
Clifty Creek Natural Area	natural bridge, Ozark headwater stream		
	oak–hickory forest, mixed bottom-land hardwoods, limestone bluff glades		
Eleven Point National Scenic River	wild stream		
Hawn State Park	oak–pine forest		
Johnson's Shut-Ins State Park	youthful stream	X	
Mark Twain National Forest	oak–hickory forest and glades	X	X
Meramec State Park	cave	X	X
	oak–hickory forest		
Mingo Wildlife Refuge	virgin vegetation		
Ozark National Scenic Riverways	wild stream		
Roaring River State Park	glades	X	X
Taberville Prairie	prairie		
MONTANA			
Anaconda–Pintlar Wilderness	wilderness areas		X
Beartooth Primitive Area	mountain zonation, alpine tundra		

Name of park or preserve	Distinctive habitats or communities	Nature trails	Information available to guide visitors
Mission Mountains Primitive Area	coniferous forest		
Selway–Bitterroot Wilderness	virgin vegetation		
Spanish Peaks Wilderness			
Bob Marshall Wilderness Glacier National Park	wilderness areas		X
	mountain zonation, alpine tundra, coniferous forest, virgin vegetation		
	wild-animal preserve		
C. M. Russell Game Range	coniferous forest, steppe		X
Cabinet Wilderness Area	wilderness area		X
	mountain zonation, coniferous forest, virgin vegetation		
Lewis and Clark Caverns State Park	cave		X
Red Rock Lakes National Refuge	wild-animal preserves		X
Salkaho Game Preserve	mountain zonation, coniferous forest, virgin vegetation		X
	wild-animal preserves		
Yellowstone National Park	hot springs, wilderness area	X	X
	mountain zonation, coniferous forest, virgin vegetation		
	wild-animal preserve		
NEBRASKA			
Chadron State Park	short-grass prairie, ponderosa pine forests		

Nebraska National Forest	man-made forest		X
NEVADA			
Desert National Wildlife Range	Sonoran desert playas		
	bighorn sheep, mule deer, pronghorn antelope		
Ruby Lake National Wildlife Refuge	marshes, ponds		
	steppe		
Sheldon National Antelope Refuge	sagebrush desert		
	bighorn sheep, pronghorn antelope, Gambel's quail, sage grouse		
Stillwater National Wildlife Refuge	alkaline desert, salt marshes, saline ponds		
NEW HAMPSHIRE			
Franconia Notch State Park	rock cliffs, talus slopes	X	X
	beech–maple forest		
Greenfield State Park	bog ponds	X	X
Hampton Beach	sand dunes		
	grasses		
Mountain Monadnock	various mountain types	X	X
Odiorne Point State Park	seashore, tidal marsh	X	X
Ossipee Lake/Heath Pond	bog pond, marshlands		
Rhododendron State Park	*Rhododendron maximum* (natural)	X	X
NEW JERSEY			
Allaire State Park	river floodplain	X	X
	holly forest		
Cheesequake State Park	salt marsh; hardwood and cedar swamps	X	X

Name of park or preserve	Distinctive habitats or communities	Nature trails	Information available to guide visitors
High Point State Park	virgin cedar swamps	X	X
Island Beach State Park	sandy beaches, dunes, mudflat, tidal marshes	X	X
	botanical preserve, holly forest, sandy upland thicket		
Lebanon, Bass River, Wharton and Pennsylvania State Forests	pine barrens	X	X
Sandy Hook State Park	tidal marsh, mudflat	X	X
Stokes State Forest	young stream	X	X
	virgin hemlock grove		
NEW MEXICO			
Bitter Lake National Wildlife Refuge	lakes, ponds, marshes		
	waterfowl		
Carlsbad Caverns National Park	cave	X	X
	desert		
San Andres National Wildlife Refuge	Chihuahuan Desert	X	X
	coniferous forest		
	bighorn sheep, mule deer		
White Sands National Monument	desert with gypsum sand dunes	X	X
NEW YORK			
Adirondack Forest Preserve	mature and youthful streams, eutrophic lake	X	
	alpine tundra, mountain zonation,		

	coniferous and beech–maple forests, sphagnum bogs, virgin vegetation		
Iroquois National Wildlife Preserve	wet meadows, swamps		
	waterfowl		
Long Island Wet Lands	salt marshes, sandy beaches		
Montezuma National Wildlife Refuge	cattail marshes		
	waterfowl		
NORTH CAROLINA			
Bladen Lakes State Forest	eutrophic lake		
	pine forests		
Blue Ridge Parkway	fir–spruce and hardwood forests, balds	X	X
Cape Hatteras National Seashore	sandy beaches, dunes		
	live oaks		
Cape Lookout National Seashore	sandy beaches, dunes		
Great Smoky Mountains National Park	youthful streams	X	X
	mountain zonation and balds, fir–spruce, beech–maple, and oak–hickory forests		
	black bear		
Pisgah National Forest	first forestry school in America	X	X
	cave, youthful stream		
	fir–spruce and hardwood forests		
Umsted State Park	hardwood forest		
Uwharrie National Forest	eutropic lake		
	hardwood and pine forest		
Wright Brothers National Memorial	dunes		

Name of park or preserve	Distinctive habitats or communities	Nature trails	Information available to guide visitors
NORTH DAKOTA			
Arrowwood National Wildlife Refuge	freshwater marshes		
	prairies		
	waterfowl, deer		
J. Clark Salyer National Wildlife Preserve	freshwater marshes, oxbow streams		
	short-grass prairie		
	waterfowl		
Lake Ilo National Wildlife Refuge	prairie grassland		
Lostwood National Wildlife Refuge	freshwater marshes, numerous potholes		
	prairies		
Snake Creek National Wildlife Refuge	lakes		
	prairies, meadows,		
	waterfowl		
Theodore Roosevelt National Memorial Park	steppe, cottonwood bottom land, sagebrush		
	buffalo herd		
OHIO			
Adams Lake Prairie Preserve	prairie	X	X
Aullwood Audubon Center			
Buzzard-Roost Rock Preserve			
Killdeer Plains Wildlife Area			
Lynx Prairie Preserve			
Resthaven Wildlife Area			

Blacklick Woods Metropolitan Park	beech–maple and swamp forests	X	X
Brown's Lake Bog Preserve	relic bog (sphagnum)		X
Cedar Bog Nature Preserve	marl bog	X	X
	wet prairie		
Clear Fork Gorge Preserve	relic forest	X	X
Cranberry Bog Preserve	relic bog (sphagnum)	X	X
Dysart Woods Preserve	oak–hickory forest	X	X
Fowler Woods Preserve	beech–maple forest	X	X
Glen Helen Nature Preserve			
Hueston Woods State Park			
Hocking Hills State Park	glacial relic forest	X	X
Mentor Marsh Nature Preserve	freshwater marsh	X	X
Oak Openings Metropolitan Park	inland sand dunes	X	X
OKLAHOMA			
Great Salt Plains Reservoir	salt flats		
	wildlife preserve		
Wichita Mountain Wildlife Preserve	prairie, scrub oak		
	buffalo and other wildlife		
OREGON			
Azalea State Park	native azaleas	X	
Booth Wayside	ponderosa pine forest		
Cape Lookout State Park	sandspit	X	
	coastal forest		
Cape Meares State Park	unique multitrunked spruce trees		
Cape Perpetua Center	sandy and rocky beaches, coastal dunes, caves		
	spruce–hemlock forest		

Name of park or preserve	Distinctive habitats or communities	Nature trails	Information available to guide visitors
Christmas Lake Lost Forest	desert, inland sand dunes ponderosa pine forest, juniper–sagebrush		
Crater Lake National Park	oligotrophic lake virgin vegetation, mountain zonation, alpine and ponderosa pine forests bird and wild-animal preserve		
Diamond Peak Wilderness Mt. Jefferson Wilderness	oligotrophic lake, youthful streams, wilderness areas mountain zonation, Douglas fir and alpine forests, virgin vegetation wild-animal preserve		
Eagle Cap Wilderness	oligotrophic lake, youthful streams, wilderness area mountain zonation, Douglas fir and alpine forests, alpine tundra, virgin vegetation wildlife preserve		
Ecola State Park	old-growth spruce–cedar–hemlock forest		
Honeyman State Park Umpqua Lighthouse State Park	coastal sand dunes	X	X

Kalmiopsis Wilderness	oligotrophic lake, youthful streams, wilderness area	
	mountain zonation, Douglas fir and ponderosa pine forests, virgin vegetation	
	wildlife preserve	
Juniper Wayside	western juniper	
Lapine Recreation Area	ponderosa pine forest	
Mary S. Young State Park	mixed coniferous–deciduous forest	
Myrtle Preserves	myrtlewood trees	
National Wildlife Refuge	eutrophic lake, freshwater marshes	
	virgin vegetation, ponderosa pine forest, juniper–sagebrush	
Nehalem Bay State Park	sandspit vegetation	
Nestucca Sandspit State Park		
Rogue River Wild and Scenic Area	youthful and wild streams	
	virgin vegetation, Douglas fir forest	
Saddle Mountain State Park	many varieties of wildflowers	X
Silver Falls State Park	typical western Oregon forest	X
South Beach State Park Yaquina Bay State Park	coastal rhododendrons	
PENNSYLVANIA		
Alan Seeger Natural Area (State Forest)	virgin white pine–hemlock–oak forest	
Algerine Swamp Natural Area	spruce–fir swamp	
Algerine Wilderness Area Hammersley Wilderness Area Wolf Run Wilderness Area	second-growth mixed oak–beech–birch–maple forests	

Name of park or preserve	Distinctive habitats or communities	Nature trails	Information available to guide visitors
Bear Meadows Natural Area (State Forest)	spruce–fir forest		
Bowman's Hill State Wildflower Preserve (State Forest)	native wildflower sanctuary	X	X
Box Huckleberry Natural Area (State Forest)	second-growth oak, spruce–fir swamp		
Bruce Lake Natural Area (State Park)	box huckleberry colony		
Bucktail State Park	second-growth oak–pine–northern hardwood forest		
Heart's Content, Allegheny National Forest	virgin white pine–hemlock forest		
Joyce Kilmer Natural Area (State Forest))			
Snyder–Middleswarth Natural Area (State Forest)			
Hemlock Natural Area (State Forest)	virgin hemlock forest		
Martin Hill Natural Area (State Forest)			
High Knob Natural Area (State Forest)	second-growth beech–birch–maple forest		
Hook Wilderness Area	second-growth mixed oak forest		
Penns Creek Wilderness Area			
Juniata Wilderness Area			
Kettle Creek Gorge Natural Area (State Forest)	second-growth beech–maple forest		
Laurel Run Natural Area	old-growth mixed oak forest		
Pine Creek Gorge Natural Area (State Forest)	second-growth mixed oak forest		
Pine Creek Natural Area	second-growth beech–birch–maple forest		
Rock Run Natural Area			
Quehanna Wilderness Area (State Forest)	second-growth mixed oak–beech–birch–maple forest		

Ricketts Glenn State Park	virgin beech–birch–maple forest		
Tionesta Natural Area, Allegheny National Forest			
RHODE ISLAND			
Arcadia Management Preserve Area	youthful stream		
Matunnuch	salt marsh, sandy beach, dunes		
Moonstone Beach			
Ninigret Conservation Area			
Seapowet Marsh			
Ocean Drive	rocky beaches		
Rockville Management Preserve Area	eutrophic lakes		
SOUTH CAROLINA			
Cape Romain National Wildlife Refuge	salt marshes, sand dunes		
Edisto Beach State Park	salt marshes, sandy beaches, dunes palmetto trees		
Hunting Island State Park			
Huntington Beach State Park			
Givhans Ferry State Park	mature stream		
Little Pee Dee State Park	inland sand dunes		
Paris Mountain State Park	youthful stream oak–hickory and pine forests	X	X
Poinsett State Park	cypress swamps	X	X
Santee State Park			
Sesquicentennial State Park	coniferous forest	X	X
Table Rock State Park	oak–hickory forest	X	X
SOUTH DAKOTA			
Custer State Park	coniferous forest bison	X	

Name of park or preserve	Distinctive habitats or communities	Nature trails	Information available to guide visitors
Jewel Cave National Monument	cave	X	X
Lake Herman State Park	eutrophic lake		
Newton Hills State Park	oak–basswood forest		
Oakwood Lakes State Park	eutrophic lake		
Sieche Hollow State Park	oak–maple–basswood forest	X	
Wind Cave National Park	cave		
	pine-studded prairie		
	wild-animal preserve, bison, prairie dogs		
TENNESSEE			
Big South Fork, Cumberland River	oak–hickory forest	X	X
Cedars of Lebanon State Forest	cedar glade flora	X	X
Cherokee National Forest	oak–hickory forest	X	X
	wild-animal preserve		
Great Smoky Mountains National Park	youthful stream, balds	X	X
	mountain zonation		
	fir–spruce, beech–maple, and oak–hickory forests		
	black bear		
Reelfoot Lake	cypress swamps	X	X
Savage Gulf	mixed mesophytic forest, distinctive vegetation	X	X

TEXAS			
Bastrop State Park	coniferous forest	X	X
Buescher State Park			
Daingerfield State Park			
Huntsville State Park			
Jim Hogg State Park			
Martin Dies, Jr., State Park			
Mission Tejas State Park			
Tyler State Park			
Bentsen-Rio Grande State Park	subtropical birdlife		
Brazos Island State Park	coastal dunes, sandy beaches		
Caddo Lake State Park	cypress swamp, coniferous forest	X	X
Galveston Island State Park	salt marsh, sandy beaches		
Longhorn Caverns State Park	cave		
Meridian State Park	golden-cheeked warbler	X	X
Monahans State Park	inland dunes		
Padre Island National Seashore	sand dunes		
Palmetto State Park	palmetto swamps	X	X
Palo Duro Canyon State Park	aoudad sheep		
Pedernales Falls State Park	golden-cheeked warbler		
UTAH			
Bear River Migratory Bird Refuge	freshwater marshes, shallow lakes		
Brown's Park National Wildlife Refuge	waterfowl		
Fish Spring National Wildlife Refuge			
Ouray National Wildlife Refuge	freshwater marshes		

Name of park or preserve	Distinctive habitats or communities	Nature trails	Information available to guide visitors
VERMONT			
Missisquoi National Wildlife Refuge	freshwater marshes, bogs	X	
VIRGINIA			
Charles C. Steirly Heron Rookery	rookery		
Goshen Pass National Area	table mountain pine, hardwood forest		
Lick Creek Natural Area	youthful stream		
Parkers Marsh Natural Area	salt marsh		
Seashore National Area	sandy beaches, dunes		
Shenandoah National Park	hardwood forest		
Wreck Island Natural Area	salt marsh, dunes		
WASHINGTON			
Columbia National Wildlife Refuge	sagebrush flats waterfowl		
Olympia National Park	rainforest		
Ridgefield National Wildlife Refuge	waterfowl		
WEST VIRGINIA			
Babcock State Park	oak–hickory and mixed hardwood forests	X	X
Blackwater Falls State Park	coniferous and mixed hardwood forests	X	X
Canaan Valley State Park	mountain zonation, alpine tundra, beech–maple and coniferous forests	X	X
Cass Scenic Railroad	mountain zonation	X	X
Cathedral State Park	virgin hemlock	X	X

Cranberry Glades	northern bog (southernmost distribution of bog rosemary)	X	X
Cranesville Swamp	northern tamarack bog (southernmost distribution)	X	X
Gaudineer Knob (Monongahela National Forest)	virgin spruce	X	X
Holly River State Park	beech–maple and mixed hardwood forests	X	X
Watoga State Park	mixed hardwood forest, hemlock, white pine	X	X
WISCONSIN			
Avoca Prairie Scientific Area	native low prairie		
Big Bay State Park	sandy beaches		
Crex Meadows Wildlife Preserve	restored prairie		X
Devil's Lake State Park	oligotrophic lake	X	X
Douglas County Grouse Area	pine and oak barrens		X
Flambeau River State Forest	wild river hemlock–hardwood forests	X	X
Goose Pond Sanctuary State Area	waterbird and shorebird preserve	X	X
Lake Winnebago	eutrophic lake	X	X
Lucius Woods State Park	virgin pine forest		
Milwaukee River Swamp State Area	mature stream		X
Peninsula Park Beech Forest	maple–beech forest		X
St. Croix River State Forest	wild river, mature stream		X
Terry Andrea State Park	Lake Michigan sand dunes		
The Ridges Sanctuary	beach ridges boreal forest	X	X
Tiffany Bottoms Wilderness Area	wilderness floodplains forest		

Name of park or preserve	Distinctive habitats or communities	Nature trails	Information available to guide visitors
WYOMING			
Bridger Wilderness Area	youthful streams, wilderness areas		
Cloud Peak Primitive Area	mountain zonation, coniferous forests, tundra, virgin vegetation		
Glacier Primitive Area			
North Absoroka Wilderness Area			
Popo Agie Primitive Area			
South Absoroka Wilderness Area			
Stratified Primitive Area			
Teton Wilderness Area			
Devil's Tower National Monument	mature stream	X	X
	steppe, coniferous forest, virgin vegetation		
	prairie dog		
Grand Teton National Park	youthful streams, wilderness areas	X	X
	mountain zonation, coniferous forest, alpine vegetation, virgin vegetation		
Yellowstone National Park	youthful streams, wilderness area	X	X
	mountain zonation, coniferous forest, alpine tundra, virgin vegetation		

Appendix 2.
Private Organizations Dealing with Environmental Problems

GENERAL

Environmental Action, Inc.
1346 Connecticut Avenue, N.W., Room 731
Washington, D.C. 20036

Friends of the Earth
529 Commercial Street
San Francisco, California 94111

Izaak Walton League of America
1326 Waukegan Road
Glenview, Illinois 60025

National Parks & Conservation Association
1710 18th Street, N.W.
Washington, D.C. 20009

Sierra Club
1050 Mill Tower
San Francisco, California 94104

CONSERVATION

American Forestry Association
1319 18th Street, N.W.
Washington, D.C. 20036

Conservation Foundation
1250 Connecticut Avenue
Washington, D.C. 20036

Defenders of Wildlife
2000 N Street, N.W.
Washington, D.C. 20036

The Desert Protective Council, Inc.
P.O. Box 33
Banning, California 92220

The Garden Club of America
Conservation Committee
598 Madison Avenue
New York, New York 10022

International Union for Conservation of Nature and Natural Resources
1110 Morges
Geneva, Switzerland

National Audubon Society
950 Third Avenue
New York, New York 10022

PUBLIC EDUCATION

Animal Welfare Institute
P.O. Box 3492
Grand Central Station
New York, New York 10017

Concern, Inc.
2233 Wisconsin Avenue, N.W.
Washington, D.C. 20007

Ecology Action Educational Institute
P.O. Box 3895
Modesto, California 95352

National Wildlife Federation
1412 Sixteenth Street, N.W.
Washington, D.C. 20036

Planned Parenthood Federation of America
810 Seventh Avenue
New York, New York 10016

Population Crisis Committee
1835 K Street, N.W.
Washington, D.C. 20006

Scientists' Institute for Public Information (SIPI)
30 E. 68th Street
New York, New York 10021

URBAN ECOLOGY

The National Urban Coalition
2100 M Street, N.W., Suite 600
Washington, D.C. 20037

The Urban Land Institute (ULI)
1200 18th Street, N.W.
Washington, D.C. 20036

OTHER

The Academy of Natural Sciences
19th and the Parkway
Philadelphia, Pennsylvania 19103

Association for Voluntary Sterilization, Inc.
14 W. 40th Street
New York, New York 10018

Consumers Education & Protective
Association International, Inc. (CEPA)
6048 Ogontz Avenue
Philadelphia, Pennsylvania 19141

National Association for Repeal of Abortion Laws
250 W. 57th Street
New York, New York 10019

Appendix 3. Federal Governmental Departments and Agencies Dealing with Environmental Problems

Bureau of Commercial Fisheries
Fish & Wildlife Service
U.S. Department of the Interior
Interior Building, C Street
Washington, D.C. 20240

Bureau of Land Management
U.S. Department of the Interior
Interior Building, C Street
Washington, D.C. 20240

Bureau of Outdoor Recreation
U.S. Department of the Interior
Interior Building, C Street
Washington, D.C. 20240

Bureau of Reclamation
U.S. Department of the Interior
Interior Building, C Street
Washington, D.C. 20240

Bureau of Sport Fisheries and Wildlife
Fish and Wildlife Service
U.S. Department of the Interior
Interior Building, C Street
Washington, D.C. 20240

Consumer Protection & Environmental Health Service
200 C Street
Washington, D.C. 20240

Environmental Control Administration
200 C Street
Washington, D.C. 20240

Environmental Science Service Administration (ESSA)
6010 Executive Boulevard
Rockville, Maryland 20852

Federal Committee on Pest Control
Agricultural Research Service
U.S. Department of Agriculture
Washington, D.C. 20204

Federal Water Pollution Control Administration
U.S. Department of the Interior
633 Indiana Avenue, N.W.
Washington, D.C. 20242

Food and Drug Administration
200 C Street
Washington, D.C. 20204

Forest Service
U.S. Department of Agriculture
Washington, D.C. 20250

Geological Survey
U.S. Department of the Interior
GSA Building
Washington, D.C. 20242

National Air Pollution Control Administration (NAPCA)
200 C Street
Washington, D.C. 20204

National Park Service
U.S. Department of the Interior
Interior Building, C Street
Washington, D.C. 20240

Plant Protection Division
U.S. Department of Agriculture
Federal Center Building
Hyattsville, Maryland 20782

Smithsonian Institution
1000 Jefferson Drive, S.W.
Washington, D.C. 20560

Soil & Water Conservation Research Division
Agricultural Research Service
U.S. Department of Agriculture
Plant Industry Station
Beltsville, Maryland 20705

Soil Conservation Service
U.S. Department of Agriculture
Washington, D.C. 20250

Urban System and Environment
U.S. Department of Transportation
800 Independence Avenue, S.W.
Washington, D.C. 20590

WORLD NATURAL VEGETATION*

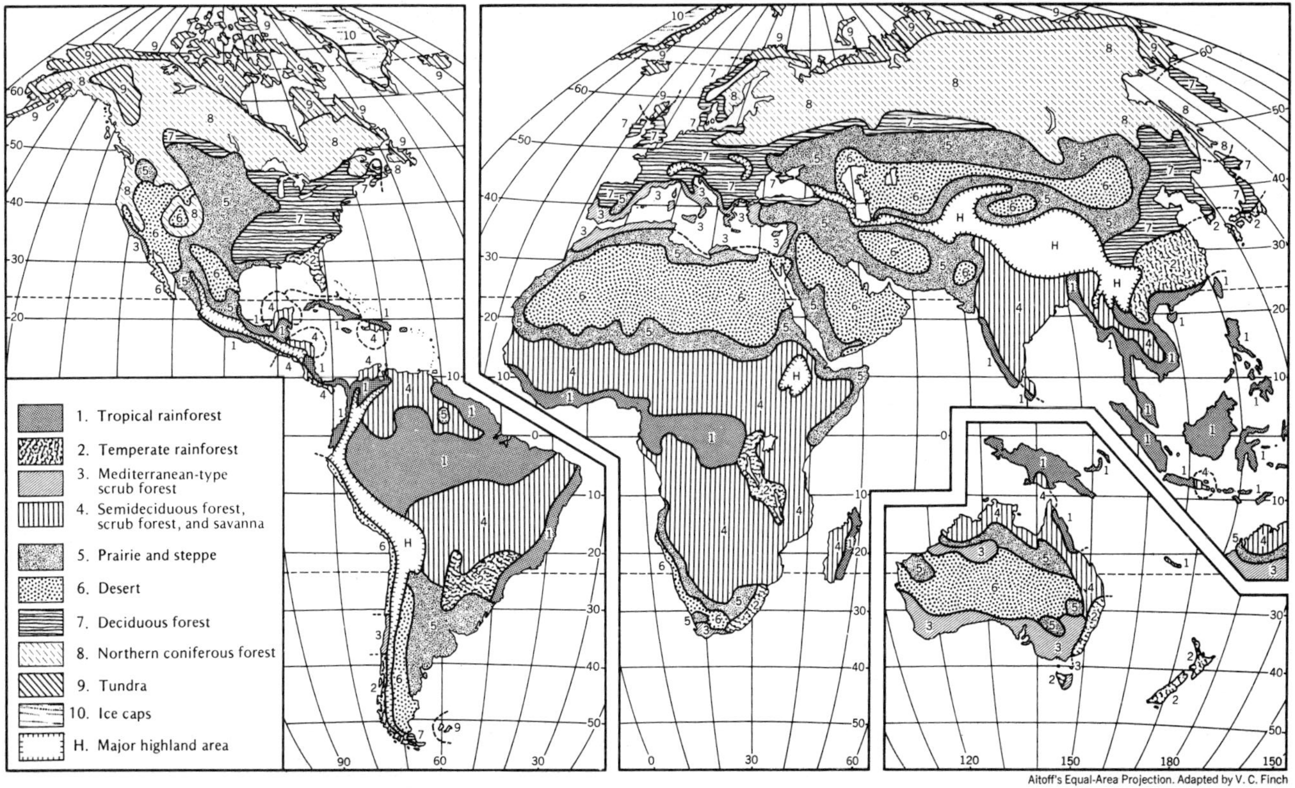

*After world map by H. Brockman Jerosch, 1951, following Rubel's classification. From *The Earth Sciences*, by Arthur N. Strahler, Harper & Row, 1971.

Index

74 75 76 77 9 8 7 6 5 4 3 2 1